THE
FIFTH
MIRACLE

The Search for the Origin
and Meaning of Life

PAUL DAVIES

SIMON & SCHUSTER PAPERBACKS
NEW YORK LONDON TORONTO SYDNEY

SIMON & SCHUSTER PAPERBACKS
Rockefeller Center
1230 Avenue of the Americas
New York, NY 10020

For information about special discounts for bulk purchases,
please contact Simon & Schuster Special Sales at
1-800-456-6798 or business@simonandschuster.com.

Designed by Ruth Lee

Manufactured in the United States of America

10 9 8 7 6 5

The Library of Congress has cataloged the hardcover edition as follows:

Davies, P. C. W.
    The fifth miracle : the search for the origin and meaning of life / Paul Davies.
        p.   cm.
    Includes bibliographical references and index.
    1. Life—Origin. 2. Life. I. Title.
    QH325.D345   1999
    576.8'3—dc21   98-33421
                                                                CIP

ISBN 978-0-684-86309-2

In memory of Keith Runcorn

CONTENTS

PREFACE

IN AUGUST 1996, the world was electrified by news that an ancient meteorite may contain evidence for life on Mars. President Clinton himself conveyed the story to the public and a startled scientific community. The momentous implications of the discovery, if such it was, were expressed in appropriate superlatives. This memorable event marked one of the few occasions when a scientific result had a direct impact on the public. Yet the plaudits and the banter glossed over the true significance of the findings.

For several years, scientists have been dramatically rethinking their ideas about the origin of life. The textbooks say that life began in some tepid pool on the Earth's surface, billions of years ago. Increasingly, however, the evidence points to a very different scenario. It now appears that the first terrestrial organisms lived deep underground, entombed within geothermally heated rocks in pressure-cooker conditions. Only later did they migrate to the surface. Astonishingly, descendants of these primordial microbes are still there, kilometers beneath our feet.

Until a few years ago, nobody suspected that life could exist in such a harsh environment, but once it was accepted that organisms can flourish beneath the Earth's surface, an even more exotic possibility presented itself. Perhaps microbes also lurk in the rocks beneath the surface of Mars? The discovery of a Martian rock containing possible fossilized bacteria was a major boost to this theory. But that was not all. Scientists were quick to spot a fascinating consequence. It could be that life actually began on Mars and then traveled to Earth in a meteorite.

The feverish excitement surrounding the Martian meteorite concealed deep divisions among the experts over the interpretation of the evidence. If confirmed, it could mean either that life has started twice in the solar system, or, alternatively, that it has spread from one planet to another. Exciting though it would be to discover that organisms can leap from planet to planet, the ultimate origin of life would still be left as an unresolved enigma if the latter explanation were correct.

How, precisely, did life begin? What physical and chemical processes can transform nonliving matter into a living organism? This much tougher problem remains one of the great scientific challenges of our age. It is currently being tackled by an army of chemists, biologists, astronomers, physicists, and mathematicians. On the basis of their research, many of them fervently conclude that the laws of nature are, to put it bluntly, rigged in favor of life. They expect that life will form wherever conditions permit—not just on Mars, but throughout the universe; and, more provocatively, in a test tube. If they are right, it will mean that life is part of the natural order of things, and that we are not alone.

Belief that life is written into the laws of nature carries a faint echo of a bygone religious age, of a universe designed for habitation by living creatures. Many scientists are scornful of such notions, insisting that the origin of life was a freak accident of chemistry, unique to Earth, and that the subsequent emergence of complex organisms, including conscious beings, is likewise purely the chance outcome of a gigantic cosmic lottery. At stake in this debate is the very place of mankind in the cosmos—who we are and where we fit into the grand scheme.

Astronomers think that the universe began in a big bang between ten and twenty billion years ago. Its explosive birth was accompanied by a flash of intense heat. During the first split second, the basic physical forces and fundamental particles of matter emerged. By the time that one second had elapsed, the essential materials of the cosmos had already formed. Space was everywhere filled with a soup of subatomic particles—protons, neutrons, and electrons—bathed in radiation at a temperature of ten billion degrees.

By present standards the universe at that epoch was astonishingly featureless. The cosmic material was spread through space with almost perfect uniformity. The temperature was the same everywhere. Matter, stripped down to its basic constituents by the fierce heat, was in a state of extraordinary simplicity. A hypothetical observer would have had no inkling from this unpromising state that the universe was primed with awesome potentialities. No clue would betray that, several billion years on, trillions of blazing stars would organize themselves into billions of spiral galaxies, that planets and crystals, clouds and oceans, mountains and glaciers would arise, that trees and bacteria and elephants and fish would inhabit one of these planets, and that this world would ring to the sound of human laughter. None of these things could be foretold.

As the universe expanded from its uniform primeval state, it cooled. And with lower temperatures came more possibilities. Matter was able to aggregate into vast amorphous structures—the seeds of today's galaxies. Atoms began to form, paving the way for chemistry and the formation of solid physical objects.

Many wonderful phenomena have emerged in the universe since that time: monstrous black holes weighing as much as a billion suns that eat stars and spew forth jets of gas; neutron stars spinning a thousand times a second, their material crushed to a billion tons per cubic centimeter; subatomic particles so elusive that they could penetrate light-years of solid lead; ghostly gravitational waves whose fleeting passage leaves no discernible imprint at all. Yet, amazing though these things may be, the phenomenon of life is more remarkable than all of them put together. It didn't bring about any sudden and dramatic alterations on the cosmic scene. In fact, if life on Earth is anything to go by, the changes it has wrought have been extremely gradual. Nevertheless, once life was initiated, the universe would never be the same. Slowly but surely it has transformed Planet Earth. And by offering a route to consciousness, intelligence, and technology, it has the potential to change the universe.

This book is about the origin of life, or biogenesis. I should state at the outset that the subject is not my professional field. I trained as a theoretical physicist. I have, however, always had a fascination for the problem of biogenesis and the related question of whether or not we are alone in the universe. I can trace my interest in these matters back to my days as a student studying physics at University College, London, in the 1960s. Like many of my friends, I read Fred Hoyle's famous science-fiction novel *The Black Cloud*, about the arrival in the solar system of a large cloud of gas from interstellar space.[1] Such clouds are well known to astronomers, but Hoyle's intriguing idea was to suppose that they could be alive. Now, this was a poser. How can a cloud be alive? I puzzled over it at length. Surely gas clouds just obey the laws of physics? How could they exhibit autonomous behavior, have thoughts, make choices? But, then, it occurred to me, all living organisms supposedly obey the laws of physics. Hoyle's brilliance was in using the example of a cloud to draw out that paradox in a stark manner.

The Black Cloud left me baffled and vaguely disturbed. What exactly, I wondered, *is* life? And how did it get started? Might there be something funny going on inside living organisms? Just at this time, my Ph.D. supervisor gave me (as an exercise in light relief) a curious paper by the highly respected physicist Eugene Wigner. The paper purported to prove that a physical system could not make a transition from a nonliving to a living state without contravening the laws of quantum physics.[2] Aha! So Wigner at least thought that something funny must have gone on when life started.

Shortly afterwards, my supervisor passed me another paper related to biology, this time by the astrophysicist Brandon Carter. It addressed an important and interesting problem concerning life, but one that dodged the need to worry about what it actually is or how it began. Carter asked the question, what properties must the physical universe have in order for life of any sort to exist at all? Suppose that by magic you could change the laws of nature or the initial conditions of the big bang. How far could you alter the basic laws or the structure of the universe, and still permit life? To take a simple example, life as we know it demands certain chemical elements, especially carbon. But few carbon atoms were made in the big bang; most were manufactured inside stars. Fred Hoyle had already noticed that the successful production of carbon in stars is actually a rather touch-and-go affair. It depends delicately on the properties of nuclear forces. Tinker with the basic laws of nuclear physics, and the universe would have little or no carbon and probably no life. Carter's ideas became known as "the anthropic principle," and suggested, audaciously, that the very existence of life is a dicey affair, a consequence of some happy coincidences in the underlying mathematical structure of the universe.

Thought-provoking though Carter's paper was, it still left the secret of life unexplained. Shortly after reading it, I got a job as a research fellow at The Institute of Theoretical Astronomy in Cambridge, where Fred Hoyle was the director and Brandon Carter a fellow researcher. During that time, I chanced across a little book by the physicist Erwin Schrödinger that seemed to address my very problem. Entitled *What Is Life?*, it set out to explain why biological organisms seem so mysterious from the point of view of physics.[3] I later discovered that this book had been immensely influential twenty years before, in the early days of the subject of molecular biology.

Unfortunately, Schrödinger's book raised more questions for me than it answered, and I consigned the problem of biogenesis to my mental "too-hard" basket. However, Carter gave me a revised copy of his paper on the anthropic principle (which he never published[4]) and, together with Bill Saslaw, another researcher at the Institute, I dabbled around with Carter's ideas. We even tried to get a meeting with Francis Crick, who at that time worked at the Medical Research Council Laboratory in Cambridge. But Crick was too busy, and Carter seemed to have pretty well sewn up the subject of the anthropic principle, so my interest in matters biological began to wane.

It was rekindled only many years later, in the early 1980s. Martin Rees (now Sir Martin Rees, the Astronomer Royal) helped organize a conference in Cambridge called "From Matter to Life." Rees, together with fellow astronomer Bernard Carr, had revitalized the subject of the anthropic principle in a famous paper published in *Nature* in 1979.[5] The conference brought together physicists and astronomers such as Brandon Carter, Freeman Dyson, and Tommy Gold, biologists like Lewis Wolpert and Sydney Brenner, mathematician John Conway, and biogenesis supremos Manfred Eigen and Graham Cairns-Smith. The agenda focused on how life began, and though no firm conclusions were drawn, the meeting served to point up the key scientific and conceptual problems. I resumed thinking about the mystery of life. Over the following decade or so, I found myself being influenced once again by the ideas of Hoyle, and also by those of Dyson and Gold. Hoyle, with collaborator Chandra Wickramasinghe, daringly suggested that maybe life did not originate on Earth at all, but was brought here by comets. Dyson also speculated on the origin of life, and let his imagination run free about the future and ultimate fate of technological civilizations. Gold had a theory that large quantities of hydrocarbons lie trapped under the ground; when a search was made to test his hypothesis, new subterranean life forms were discovered. All these developments helped shape my thinking on the subject.

Another person who greatly influenced my interest in biogenesis was the late Keith Runcorn, my former colleague at the University of Newcastle upon Tyne. Runcorn was a geophysicist whose interests extended well beyond Earth into the solar system. Although geophysics was far from my own area of expertise, I would often sit in on Keith's seminars and conferences. The fiftieth meeting

of the Meteoritical Society, held in Newcastle in 1987, was especially memorable, for it was there that I first learned about the Martian meteorites.

The final piece of the jigsaw came in the early 1990s, by which
time I had moved to Australia to work at The University of Adelaide. There I became interested in the work of Duncan Steel, an
expert on asteroid and cometary impacts with the planets. It was
Steel who introduced me to the fact that material could be ejected
from the planets by cosmic collisions, an idea that laid the foundations for my theory about micro-organisms traveling between Mars
and Earth.

When I set out to write this book, I was convinced that science
was close to wrapping up the mystery of life's origin. The dramatic
evidence for microbes living deep underground promised to provide
the "missing link" between the prebiotic world of biochemical soups
and the first primitive cells. And it is true that many scientists working in this field confidently believe that the major problems of biogenesis have largely been solved. Several recent books convey the
confident message that life's origin is not really so mysterious after
all.[6] However, I think they are wrong. Having spent a year or two re-
searching the field, I am now of the opinion that there remains a
huge gulf in our understanding. To be sure, we have a good idea of
the where and the when of life's origin, but we are a very long way
from comprehending the how.

This gulf in understanding is not merely ignorance about certain technical details, it is a major conceptual lacuna. I am not suggesting that life's origin was a supernatural event, only that we are
missing something very fundamental about the whole business. If it
is the case, as so many experts and commentators suggest, that life is
bound to arise given the right conditions, then something truly
amazing is happening in the universe, something with profound
philosophical ramifications. My personal belief, for what it is worth,
is that a fully satisfactory theory of the origin of life demands some
radically new ideas.

Many investigators feel uneasy about stating in public that the origin of life is a mystery, even though behind closed doors they freely admit that they are baffled. There seem to be two reasons for their unease. First, they feel it opens the door to religious fundamentalists and their god-of-the-gaps pseudo-explanations.[7] Second, they worry that a frank admission of ignorance will undermine funding, especially for the search for life in space. The view seems to be that governments are more likely to spend money seeking extraterrestrial life if scientists are already convinced that it is out there.

In my opinion, this attitude is totally misguided. Scientists do their disciplines no credit by making exaggerated claims merely for public consumption. More important, ignorance provides a much better motivation for experiment than certainty. It is important to seek life on other worlds, and to try and synthesize it in the laboratory, precisely *because* we are so uncertain of how it came to be. For this reason, I strongly support NASA's new Astrobiology program. If I am right that biogenesis hints at something profoundly new and amazing, then searching other worlds may enable us to catch this remarkable transition in the act. Astronomers consider the outer planets like Saturn and Jupiter and their moons to be gigantic prebiotic laboratories, where the steps that led to life on Earth have been frozen in time, poised partway between the realm of complex chemistry and the realm of true biology.

In the case of Mars, it seems likely that the line between nonlife and life will have been crossed, and that, at some stage in the past, life flourished on the red planet. In fact, for reasons I shall explain in this book, I believe that past life on Mars was a virtual certainty. I also think there is a good chance of finding life there today, if you know where to look.

Solving the mystery of biogenesis is not just another problem on a long list of must-do scientific projects. Like the origin of the universe and the origin of consciousness, it represents something altogether deeper, because it tests the very foundations of our science and our world-view. A discovery that promises to change the very principles on which our understanding of the physical world is built deserves to be treated as an urgent priority. The mystery of life's origin has puzzled philosophers, theologians, and scientists for over two and a half millennia. During the next decade, we have a golden opportunity to make some major advances in this field. That scientists are currently stumped makes this opportunity all the more exciting and compelling.

I believe that we will not solve the problem of biogenesis without first having a deep understanding of the nature of life. What, exactly, is it? Life is so extraordinary in its properties that it qualifies for the description of an alternative state of matter. I begin the book by seeking a definition of life—a notoriously difficult problem. Most textbooks focus on the chemistry of life: which molecules do what inside the cell. Obviously, life is a chemical phenomenon, but its distinctiveness lies not in the chemistry as such. The secret of life comes instead from its informational properties; a living organism is a complex information-processing system.

Complexity and information can be illuminated by the subject of thermodynamics, a branch of science that links physics, chemistry, and computation. For decades there has been a suspicion that life is so amazing that it must somehow circumvent the laws of thermodynamics. In particular, the second law of thermodynamics, arguably the most fundamental of all the laws of nature, describes a trend of decay and degeneration that life clearly bucks. I have devoted chapter 2 to an extensive discussion of the second law of thermodynamics, for it provides the context for what I regard as the ultimate problem of biogenesis: namely, where biological information came from. Whatever remarkable chemistry may have occurred on the primeval Earth or some other planet, life was sparked not by a molecular maelstrom as such, but—somehow!—by *the organization of information*. It is a theme I develop further in chapters 3, 4, and 5, where I describe the various competing theories of primordial soups and other scenarios for turning chemistry into life, plus some of the attempts to create life in the laboratory. I also give a brief review of the fossil evidence for the earliest life forms. Some of the introductory sections on Darwinism and basic molecular biology may be familiar to the reader, and could be skipped. However, I have tried to present the orthodox ideas with a novel slant.

If I am right that the key to biogenesis lies, not with chemistry, but with the formation of a particular logical and informational architecture, then the crucial step involved the creation of an information-processing system, employing software control. In chapter 4, I argue that this step was closely associated with the appearance of the genetic code. Bringing some of the language and concepts of computation to the problem, I have endeavored to throw light on the highly novel form of complexity that is found in the genes of living organisms. The peculiarity of biological complexity make genes seem almost like impossible objects—yet they must have formed somehow. I have come to the conclusion that no familiar law of nature could produce such a structure from incoherent chemicals with the inevitability that some scientists assert. If life does form easily, and is common throughout the universe, then new physical principles must be at work. It is a theme that I take up in the final chapter, where I have tried to spell out the immense philosophical ramifications that follow if the universe teems with life—as many people seem to believe is the case. Though I have no doubt that the origin of life was not in fact a miracle, I do believe that we live in a bio-friendly universe of a stunningly ingenious character.

Most of the latter half of the book is devoted to a radical new theory for the origin of life. Since the time of Darwin, there have been only two broad theories of biogenesis. The first is that life began by chemical self-assembly in a watery medium somewhere on the Earth's surface—Darwin himself wrote of a "warm little pond."[8] The other is that life came to Earth from space in the form of already viable microbes—the so-called panspermia hypothesis. In the latter scenario, the ultimate origin of life is left as a mystery. In recent years, however, the evidence has increasingly suggested to me a third alternative: that life began *inside* the Earth. Not in the far interior, of course, but several kilometers down in the solid crust, probably beneath the seabed, where geothermal activity creates cauldronlike conditions. The extreme heat and chemical potency of the subsurface zone, especially near volcanic vents, would instantly kill most known organisms. However, such an environment was ideal for biogenesis, and scientists have discovered bizarre microbes still living in these scalding locations today, at temperatures well above the boiling point of water. These superbugs are described in chapter 7, where I argue that they are living fossils left over from the dawn of life.

I believe that very similar superbugs once lived beneath the surface of Mars, and may well exist there today, far underground, for reasons explained in chapter 8. Furthermore, I am convinced that micro-organisms have traveled between Earth and Mars inside rocks blasted from these planets by the impacts of giant meteorites. A large part of chapter 8 is devoted to the contentious issue of Martian meteorites, especially the famous ALH84001, which NASA scientists have claimed contains fossil Martian microbes. The near-certainty of planetary cross-contamination, which seems to have been overlooked by most scientists and commentators involved in the recent life-on-Mars debate, makes the ultimate origin of life problematic. Did it start on Earth, Mars, or both independently? Or somewhere else entirely? I discuss the importance of astronomy to biogenesis in chapter 6, and review the evidence for revived panspermia theories in chapter 9.

In preparing this book, I have benefited considerably from detailed discussions with many distinguished colleagues. Some I have already mentioned. Special thanks must go to Susan Barns, Robert Hannaford, John Parkes, Steven Rose, Mike Russell, Duncan Steel, and Malcolm Walter, all of whom kindly read and commented on early drafts of the manuscript. Other people who have given me valuable assistance during the writing phase are Diane Addie, David Blair, Julian Brown, Roger Buick, Julian Chela-Flores, George Coyne, Helena Cronin, Robert Crotty, Susan Davies, Reza Ghadiri, Monica Grady, Gerry Joyce, Stuart Kauffman, Bernd-Olaf Küppers, Clifford Matthews, Chris McKay, Jay Melosh, Curt Mileikowsky, Martin Redfern, Martin Rees, Everett Shock, Lee Smolin, Karl Stetter, Roger Summons, Ruediger Vaas, Frances Westall, and Ian Wright.

Finally, I should like to add a few words about the title of this book. It derives from the biblical account in the book of Genesis, which describes how God made the world in a series of specific steps. Verse 11 of the first chapter states, "Let the land produce vegetation." This is the first mention of life, and it seems to be the fifth miracle. The preceding four miracles are the creation of the universe, the creation of light, the creation of the firmament, and the creation of dry land. Biblical scholars tell me that this enumeration is a misreading of Genesis, because the opening line, "In the beginning God created the heavens and the earth," is not in fact the description of a miraculous act, but a statement of the overall agenda that is itemized in the subsequent verses. Nevertheless, I have stuck to the fifth miracle. In using this title, I am not suggesting that the origin of life actually was a miracle. I refer those readers interested in the theological aspects of this topic to my earlier books *The Mind of God*[9] and *Are We Alone?*[10]

PAUL DAVIES
Adelaide, South Australia

THE FIFTH MIRACLE

CHAPTER I

The Meaning of Life

IMAGINE BOARDING A TIME MACHINE and being transported back four billion years. What will await you when you step out? No green hills or sandy shores. No white cliffs or dense forests. The young planet bears little resemblance to its equable appearance today. Indeed, the name "Earth" seems a serious misnomer. "Ocean" would suit better, for the whole world is almost completely submerged beneath a deep layer of hot water. No continents divide the scalding seas. Here and there the peak of a mighty volcano thrusts above the surface of the water and belches forth immense clouds of noxious gas. The atmosphere is crushingly dense and completely unbreathable. The sky, when free of cloud, is lit by a sun as deadly as a nuclear reactor, drenching the planet in ultraviolet rays. At night, bright meteors flash across the heavens. Occasionally a large meteorite penetrates the atmosphere and plunges into the ocean, raising gigantic tsunamis, kilometers high, which crash around the globe.

The seabed at the base of the global ocean is unlike the familiar rock of today. A Hadean furnace lies just beneath, still aglow with primeval heat. In places the thin crust ruptures, producing vast fissures from which molten lava erupts to invade the ocean depths. The seawater, prevented from boiling by the enormous pressure of the overlying layers, infuses the labyrinthine fumaroles, creating a tumultuous chemical imbroglio that reaches deep into the heaving crust. And somewhere in those torrid depths, in the dark recesses of the seabed, something extraordinary is happening, something that is destined to reshape the planet and, eventually perhaps, the universe. Life is being born.

The foregoing description is undeniably a speculative reconstruction. It is but one of many possible scenarios offered by scientists for the origin of life, but increasingly it seems the most plausible. Twenty years ago, it would have been heresy to suggest that life on Earth began in the torrid volcanic depths, far from air and sunlight. Yet the evidence is mounting that our oldest ancestors did not crawl out of the slime so much as ascend from the sulfurous underworld. It may even be that we surface dwellers are something of an aberration, an eccentric adaptation that arose only because of the rather special circumstances of Earth. If there is life elsewhere in the universe, it may well be almost entirely subterranean, and only rarely manifested on a planetary surface.

Although there is now a measure of agreement that Earth's earliest bioforms were deep-living microbes, opinion remains divided over whether life actually began way down in the Earth's crust, or merely took up residence there early on. For, in spite of spectacular progress over the past few decades in molecular biology and biochemistry, scientists still don't know for sure how life began. The outline of a theory is available, but we are a long way from having a blow-by-blow account of the processes that transformed matter into life. Even the exact location of the incubator remains a frustrating mystery. It could be that life didn't originate on Earth at all; it may have come here from space.

The challenge facing scientists struggling to explain the origin of life is the need to piece together a narrative of events that happened billions of years ago and have left little or no trace. The task is a daunting one. Fortunately, during the last few years some remarkable discoveries have been made about the likely nature of Earth's most primitive organisms. There have also been great strides in laboratory procedures, and a growing understanding of conditions in the early solar system. The recent revival of interest in the possibility of life on Mars has also served to broaden the thinking about the conditions necessary for life. Together, these developments have elevated the subject from a speculative backwater of science to a mainstream research project.

The problem of how and where life began is one of the great outstanding mysteries of science. But it is more than that. The story of life's origin has ramifications for philosophy and even religion. Answers to such profound questions as whether we are the only sentient beings in the universe, whether life is the product of random accident or deeply rooted law, and whether there may be some sort of ultimate meaning to our existence, hinge on what science can reveal about the formation of life.

In a subject supercharged with such significance, lack of agreement is unsurprising. Some scientists regard life as a bizarre chemical freak, unique in the universe, whereas others insist that it is the expected product of felicitous natural laws. If the magnificent edifice of life is the consequence of a random and purely incidental quirk of fate, as the French biologist Jacques Monod claimed, we must surely find common cause with his bleak atheism, so eloquently expressed in these words: "The ancient covenant is in pieces: man at last knows that he is alone in the unfeeling immensity of the universe, out of which he has emerged only by chance. Neither his destiny nor his duty have been written down."[1] But if it transpires that life emerged more or less on cue as part of the deep lawfulness of the cosmos—if it is scripted into the great cosmic drama in a basic manner—it hints at a universe with a purpose. In short, the origin of life is the key to the meaning of life.

In the coming chapters I shall carefully examine the latest scientific evidence in an attempt to confront these contentious philosophical issues. Just how bio-friendly is the universe? Is life unique to Planet Earth? How can something as complex as even the simplest organism be the product of straightforward physical processes?

Life's mysterious origin

> The origin of life appears . . . to be almost a miracle, so
> many are the conditions which would have had to be
> satisfied to get it going.
>
> FRANCIS CRICK[2]

According to the Australian Aborigines of the Kimberley, in the Creation Time of Lalai, Wallanganda, the sovereign of the galaxy and maker of the Earth, let fresh water fall from space upon Wunggud, the giant Earth Snake. Wunggud, whose very body is made of the primeval material, was coiled into a ball of jellylike substance, *ngallalla yawun*. On receiving the invigorating water, Wunggud stirred. She formed depressions in the ground, *garagi*, to collect the water. Then she made the rain, and initiated the rhythmic processes of life: the seasons, the reproductive cycles, menstruation. Her creative powers shaped the landscape and brought forth all creatures and growing things, over which she still holds dominion.[3]

All cultures have their creation myths, some more colorful than others. For centuries, Western civilization looked to the Bible for enlightenment on the subject. The biblical text seems disappointingly bland when set beside the Australian story: God created life in more or less its present form *ab initio*, as the fifth miracle.

Not far from the Kimberley—across the Great Sandy Desert, in the mountains of the Pilbara—lie the oldest known fossils on Earth. These extraordinary remains form part of the scientific account of creation. Science takes as its starting point the assumption that life wasn't made by a god or a supernatural being: it happened unaided and spontaneously, as a natural process.

Over the past two centuries, scientists have painstakingly pieced together the history of life. The fossil record shows clearly that ancient life was very different from extant life. Generally speaking, the farther back in time you go, the simpler were the living things that inhabited Earth. The great proliferation of complex life forms occurred only within the last billion years. The oldest well-documented true animal fossils, also to be found in Australia (in the Flinders Ranges, north of Adelaide), are dated at 560 million years. Known as Ediacara, they include creatures resembling jellyfish. Shortly after this epoch, about 545 million years ago, there began a veritable explosion of species, culminating in the colonization of the land by large plants and animals. But before about one billion years ago, life was restricted to single-celled organisms. This record of complexification and diversification is broadly explained by Darwin's theory of evolution, which paints a picture of species continually branching and rebranching to form more and more distinct lineages. Conversely, in the past these lineages converge. The evidence strongly affirms that all life on Earth descended via this branching process from a common ancestor. That is, every person, every animal and plant, every invisible bacterium can be traced back to the same tiny microbe that lived billions of years ago, and thence back to the first living thing.[4] What remains to be explained—what stands out as the central unsolved puzzle in the scientific account of life—is how the first microbe came to exist.

Peering into life's innermost workings serves only to deepen the mystery. The living cell is the most complex system of its size known to mankind. Its host of specialized molecules, many found nowhere else but within living material, are themselves already enormously complex. They execute a dance of exquisite fidelity, orchestrated with breathtaking precision. Vastly more elaborate than the most complicated ballet, the dance of life encompasses countless molecular performers in synergetic coordination. Yet this is a dance with no sign of a choreographer. No intelligent supervisor, no mystic force, no conscious controlling agency swings the molecules into place at the right time, chooses the appropriate players, closes the links, uncouples the partners, moves them on. The dance of life is spontaneous, self-sustaining, and self-creating.

How did something so immensely complicated, so finessed, so exquisitely clever, come into being all on its own? How can mindless molecules, capable only of pushing and pulling their immediate neighbors, cooperate to form and sustain something as ingenious as a living organism?

Solving this riddle is an exercise in many disciplines—biology foremost, but chemistry, geology, astronomy, mathematics, computing, and physics contribute too. It is also an exercise in history. Few scientists believe that life began in a single monumental leap. No physical process abruptly "breathed life" into inert matter. There must have been a long and complicated transitional stage between the nonliving and the first truly living thing, an extended chronology of events unlikely to be preordained in its myriad details. A law of nature could not alone explain how life began, because no conceivable law would compel a legion of atoms to follow precisely a prescribed sequence of assemblage. So, although complying with the laws of nature, the actual route to life must have owed much to chance and circumstance—or contingency, as philosophers call it. Because of this, and because of our ignorance about the conditions that prevailed in the remote past, we will never know exactly which particular sequence of events produced the first life form.

The mystery of biogenesis runs far deeper than ignorance over details, however. There is also a profound conceptual problem concerning the very nature of life. I have on my desk one of those lamps, popular in the 1960s, containing two differently colored fluids that don't mix. Blobs of one fluid slowly rise and fall through the other. People often comment that the behavior of the blobs is "lifelike." The lamp is not alone in this respect. Many inanimate systems have lifelike qualities—flickering flames, snowflakes, cloud patterns, swirling eddies in a river. What is it that distinguishes genuine living organisms from merely lifelike systems? It is not simply a matter of degree; there is a real difference between the nature of the living and the merely lifelike. If a chicken lays an egg, it is a fair bet that the hatched fledgling will also be a chicken; but try predicting the precise shape of the next snowflake. The crucial difference is that the chicken is made according to specific genetic instructions, whereas lamp blobs, snowflakes, and eddies form willy-nilly. There is no gene for a snowflake. Biological complexity is *instructed* complexity or, to use modern parlance, it is information-based complexity. In the coming chapters I shall argue that it is not enough to know how life's immense structural complexity arose; we must also account for the origin of biological information. As we shall see, scientists are still very far from solving this fundamental conceptual puzzle. Some people rejoice in such ignorance, imagining that it leaves room for a miraculous creation. However, it is the job of science to solve mysteries without recourse to divine intervention. Just because scientists are still uncertain how life began does not mean life cannot have had a natural origin.

How does one go about assembling a scientific account of the genesis of life? At first sight the task seems hopeless. The traditional method of seeking rock fossils offers few clues. Most of the delicate prebiotic molecules that gave rise to life will long ago have been eradicated. The best we can hope for is some degraded chemical residue of the ancestral organisms from which familiar cellular life evolved.

If we had to rely on rock fossils alone, the task of understanding the origin and early evolution of life would indeed be formidable. Fortunately, there is another line of evidence altogether. It too stretches back into the dim and distant past, but it exists right here and now, inside extant life forms. Biologists are convinced that relics of ancient organisms live on in the structures and biochemical processes of their descendants—including human beings. By studying how the modern cell operates, we can glimpse remnants of ancestral life at work—a peculiar molecule here, an odd chemical reaction there—in the same way that out-of-place coins, rusty tools, or suspicious mounds of earth alert the archaeologist. So, amid the intricate processes going on inside modern organisms, traces of primeval life survive, forming a bridge with our distant past. Analyzing these obscure traces, scientists have made a start on reconstructing the physical and chemical pathways that may have brought the first living cell into existence.

Even with such biochemical clues, the task of reconstruction would still be largely guesswork were it not for the recent discovery of certain "living fossils"—microbes that inhabit weird and extreme environments. These so-called superbugs are being intensively investigated, and look set to revolutionize microbiology. It could be that we are glimpsing in these offbeat microbes something close to the primitive organisms that spawned all life on Earth. More clues may come from the search for life on Mars and other planets, and the study of comets and meteorites. By piecing together all these strands of evidence, we may yet be able to deduce, in broad outline at least, the way in which life first emerged in the universe.

What is life?

Before we tackle the problem of its origin, it is important to have a clear idea of what life is. Fifty years ago, many scientists were convinced the mystery of life was about to be solved. Biologists recognized that the key lay among the molecular components within the cell. Physicists had by then made impressive strides elucidating the structure of matter at the atomic level, and it looked as if they would soon clear up the problem of life too. The agenda was set by the publication of Erwin Schrödinger's book *What Is Life?* in 1944. Living organisms, it seemed at the time, would turn out to be nothing more than elaborate machines with microscopic parts that could be studied using the techniques of experimental physics. Careful investigation lent support to this view. The living cell is indeed crammed with miniature machines. All it required was an assembly manual and the problem would be solved. Today, however, the picture of the cell as nothing but a very complicated mechanism seems rather naïve. To be sure, molecular biology has scored some dazzling successes, but scientists still can't quite put their finger on *exactly* what it is that separates a living organism from other types of physical objects. Though treating organisms as mechanisms has undoubtedly proved very fruitful, it is important not to be mesmerized by its simplistic charm. Mechanistic explanation is an important part of understanding life, not the whole story.

Let me give a striking example of where the problem lies. Imagine throwing a dead bird and a live bird into the air. The dead bird will land with a thud, predictably, a few meters away. The live bird may well end up perched improbably on a television aerial across town, on the branch of a tree, on a rooftop, in a hedgerow, or in a nest. It would be hard to guess in advance exactly where.

As a physicist, I am used to thinking of matter as passive, inert and clodlike, responding only when coerced by external forces—as when the dead bird plunges to the ground under the tug of gravity. But living creatures literally have a life of their own. It is as if they contain some inner spark that gives them autonomy, so that they can (within limits) do as they please. Even bacteria do their own thing in a restricted way. Does this inner freedom, this spontaneity, imply that life defies the laws of physics, or do organisms merely harness those laws for their own ends? If so, how? And where do such "ends" come from in a world apparently ruled by blind and purposeless forces?

This property of autonomy, or self-determination, seems to touch on the most enigmatic aspect that distinguishes living from nonliving things, but it is hard to know where it comes from. What physical properties of living organisms confer autonomy upon them? Nobody knows.

Autonomy is one important characteristic of life. But there are many others, including the following:

Reproduction. A living organism should be able to reproduce. However, some nonliving things, like crystals and bush fires, can reproduce, whereas viruses, which many people would regard as living, are unable to multiply on their own. Mules are certainly living, even though, being sterile, they cannot reproduce. A successful offspring is more than a mere facsimile of the original; it also includes *a copy of the replication apparatus*. To propagate their genes beyond the next generation, organisms must replicate the means of replication, as well as replicating the genes themselves.

Metabolism. To be considered as properly alive, an organism has to *do* something. Every organism processes chemicals through complicated sequences of reactions, and as a result garners energy to enable it to carry out tasks, such as movement and reproduction. This chemical processing and energy liberation is called metabolism. However, metabolism cannot be equated with life. Some micro-organisms can become completely dormant for long periods of time, with their vital functions shut down. We would be reluctant to pronounce them dead if it is possible for them to be revived.

Nutrition. This is closely related to metabolism. Seal up a living organism in a box for long enough and in due course it will cease to function and eventually die. Crucial to life is a continual throughput of matter and energy. For example, animals eat, plants photosynthesize. But a flow of matter and energy alone fails to capture the real business of life. The Great Red Spot of Jupiter is a fluid vortex sustained by a flow of matter and energy. Nobody suggests it is alive. In addition, it is not energy as such that life needs, but something like useful, or free, energy. More on this later.

Complexity. All known forms of life are amazingly complex. Even single-celled organisms such as bacteria are veritable beehives of activity involving millions of components. In part, it is this complexity that guarantees the unpredictability of organisms. On the other hand, a hurricane and a galaxy are also very complex. Hurricanes are notoriously unpredictable. Many nonliving physical systems are what scientists call chaotic— their behavior is too complicated to predict, and may even be random.

Organization. Maybe it is not complexity *per se* that is significant, but *organized* complexity. The components of an organism must cooperate with each other or the organism will cease to function as a coherent unity. For example, a set of arteries and veins are not much use without a heart to pump blood through them. A pair of legs will offer little locomotive advantage if each leg moves on its own, without reference to the other. Even within individual cells the degree of cooperation is astonishing. Molecules don't simply career about haphazardly, but show all the hallmarks of a factory assembly line, with a high degree of specialization, a division of labor, and a command-and-control structure.

Growth and development. Individual organisms grow and ecosystems tend to spread (if conditions are right). But many nonliving things grow too (crystals, rust, clouds). A subtler yet altogether more significant property of living things, treated as a class, is development. The remarkable story of life on Earth is one of gradual evolutionary adaptation, as a result of variety and novelty. Variation is the key. It is replication combined with variation that leads to Darwinian evolution. We might consider turning the problem upside down and say: if it evolves in the way Darwin described, it lives.

Information content. In recent years scientists have stressed the analogy between living organisms and computers. Crucially, the information needed to replicate an organism is passed on in the genes from parent to offspring. So life is information technology writ small. But, again, information as such is not enough. Though there is information aplenty in the positions of the fallen leaves in a forest, it doesn't *mean* anything. To qualify for the description of living, information must be meaningful to the system that receives it: there must be a "context." In other words, the information must be *specified*. But where does this context itself come from, and how does a meaningful specification arise spontaneously in nature?

Hardware/software entanglement. As we shall see, all life of the sort found on Earth stems from a deal struck between two very different classes of molecules: nucleic acids and proteins. These groups complement each other in terms of their chemical properties, but the contract goes much deeper than that, to the very heart of what is meant by life. Nucleic acids store life's software; the proteins are the real workers and constitute the hardware. The two chemical realms can support each other only because there is a highly specific and refined communication channel between them mediated by a code, the so-called genetic code. This code, and the communication channel—both advanced products of evolution—have the effect of entangling the hardware and software aspects of life in a baffling and almost paradoxical manner.

Permanence and change. A further paradox of life concerns the strange conjunction of permanence and change. This ancient puzzle is sometimes referred to by philosophers as the problem of being versus becoming. The job of genes is to replicate, to conserve the genetic message. But without variation, adaptation is impossible and the genes will eventually get snuffed out: adapt or die is the Darwinian imperative. How do conservation and change coexist in one system? This contradiction lies at the heart of biology. Life flourishes on Earth because of the creative tension that exists between these conflicting demands; we still do not fully understand how the game is played out.

It will be obvious that there is no easy answer to Schrödinger's question: what is life? No simple defining quality distinguishes the living from the nonliving. Perhaps that is just as well, because science presents the natural world as a unity. Anything that drives a wedge between the domains of the living and the nonliving risks biasing us towards the belief that life is magical or mystical, rather than something entirely natural. It is a mistake to seek a sharp dividing line between living and nonliving systems. You can't strip away the frills and identify some irreducible core of life, such as a particular molecule. There is no such thing as a living molecule, only a system of molecular processes that, taken collectively, may be considered alive.

I can summarize this list of qualities by stating that, broadly speaking, life seems to involve two crucial factors: metabolism and reproduction. We can see that in our own lives. The most basic things that human beings do are breathe, eat, drink, excrete, and have sex. The first four activities are necessary for metabolism; the last is necessary for reproduction. It is doubtful that we would consider a population of entities that have metabolism but no reproduction, or reproduction without metabolism, to be living in the full sense of the term.

The life force and other discredited notions

Given the elusive character of life, it is not surprising that some people have resorted to mystical interpretations. Perhaps organisms are infused with some sort of essence or soul that brings them alive? The belief that life requires an extra ingredient over and above ordinary matter obeying normal physical laws is known as vitalism. It is a beguiling idea with a long history. The Greek philosopher Aristotle proposed that a special quality which he called the life force, or psyche, bestowed upon living organisms their remarkable properties, especially that of autonomy or self-movement. Aristotle's psyche was different from the later Christian idea of the soul as a special and separate entity. Indeed, in Aristotle's scheme, everything in the universe was considered to possess intrinsic properties that determined its behavior. In effect, he regarded the whole cosmos as an organism.

Over the centuries, the notion of a life force reappeared in many different guises. From time to time attempts were made to link it with specific substances—for example, air. Perhaps this was not unreasonable; after all, breathing stops on death, and artificial respiration can sometimes restore vital functions. Later, blood became the life-giving substance. These ancient myths live on in expressions like "breathing life" into something, or "draining away the lifeblood," as if there were more than one kind of blood.

As scientific understanding advanced, so the life force became associated with more sophisticated concepts. Claims were made that it was attributable to phlogiston, or the ether—imaginary substances that themselves became discredited in due course. Another idea, popular in the eighteenth century, was to identify the life force with electricity. At that time electrical phenomena were sufficiently mysterious to serve such a purpose, and Volta's famous experiments demonstrated that electricity could make severed frog muscles twitch. The belief that electricity could revivify matter was dramatically exploited by Mary Shelley in her famous novel *Frankenstein*, in which the monster, assembled from dead human organs, is brought to life with a huge spark from a thunderstorm. In the late nineteenth century, radioactivity replaced electricity as the latest mysterious phenomenon; sure enough, claims were made that a solution of gelatine could be instilled with life by exposing it to emissions from radium crystals.

These early attempts to pin down the life force appear to us today as plain daft. Nevertheless, the assumption that life requires something in addition to normal physical forces survived well into the twentieth century. For a long time, chemicals made by organisms were regarded as somehow different from the rest. Even today, the subject of chemistry is divided into "organic" and "inorganic." The implication was that organic substances like alcohol, formaldehyde, and urea somehow retain the magical essence of life even when separated from any living organism. By contrast, inorganic substances such as common salt are well and truly dead.

It came as something of a shock to vitalists when, in 1828, Friedrich Wöhler managed to synthesize urea from ammonium cyanate, an inorganic substance. By breaching the invisible barrier between the inorganic and organic worlds, and demonstrating that life itself was not needed to make organic substances, Wöhler scotched the idea that organic chemicals are subtly different from the rest. No longer was it necessary to posit two distinct types of matter. A common set of principles would henceforth govern the chemistry of both the living and the nonliving world. We now know that atoms are cycled through the biosphere, in and out of living organisms, all the time. Every carbon atom in your body is identical to a carbon atom in the air or in a lump of chalk. There is no mysterious "zing" that renders your carbon atoms "alive" while those around you are dead; no lifelike quality that a carbon atom acquires when you eat it, and gives up when you exhale it.

In spite of the blurring of the distinction between organic and inorganic chemistry, vitalism lived on, popularized by some well-known philosophers such as Henri Bergson in France. In fact, it entered a more scientific phase with the work of a German embryologist, Hans Driesch, in the early 1900s. Driesch was impressed that embryos could be mutilated early in their growth yet still recover to produce a normal organism. These and other remarkable properties of organic development led him to propose that the emergence of the correct form of the organism, in all its intricate complexity, must be under the control of a guiding life force, which he termed entelechy. Driesch realized that the ordering properties of entelechy would place it in conflict with normal physical forces and the law of conservation of energy. He suggested that entelechy operates by affecting the timing of molecular interactions in a way that introduces a cooperative, holistic pattern.

Although embryo development remains incompletely understood, enough is known about it, and biological pattern formation in general, to convince biologists that entelechy, like any other version of the life-force concept, is an unnecessary complication. This hasn't prevented many nonscientists from clinging to vitalistic ideas today. Beliefs range from the quasi-scientific, such as Kirlian photography, where a photographic image showing a sort of corona glow around a person's hand is produced by placing it in a strong electric field, to the unashamedly mystical ideas of yin and yang energy flows, karmas, and auras that appear only to gifted psychics. Unfortunately for the mystics, no properly conducted scientific experiment has ever demonstrated a life force at work, nor do we need such a force to explain what goes on inside biological organisms.

A further reason to reject vitalistic explanations of life is their totally *ad hoc* character. If the life force manifests itself only in living things, it has little or no explanatory value. To make this point clear, let me use the analogy of a steam locomotive. Ask: what is a steam locomotive and how does it work? An engineer could give a very detailed reply to this question. He could tell you about pistons and governors and steam pressure and the thermodynamics of combustion. He could say which bits moved what to make the wheels turn. He might also wax lyrical and describe the gleaming brass and belching smoke.

Now, it might be objected that the engineer's account, however complete, would still leave out the essential *traininess* of the locomotive, the thing that endows a mere heap of connected metallic parts with the thrilling power, the majesty, the elegance of movement, the sense of presence that one associates with a steam locomotive. So are we to suppose that, in addition to being a collection of metal components, a locomotive must also be infused with "traininess" to make it the genuine item?

Of course, that is absurd. Where else are we to find traininess other than in a train? The steam locomotive simply *is* the bits and pieces of which it is composed, arranged together in the manner that they are. That is all. There is no extra ingredient, no traininess, that the manufacturer must add to "bring the machine alive" for its intended function. Likewise, in seeking to understand the origin of life, scientists look to normal molecular processes to explain what happened, and not to an external life force to enliven dead matter. What makes life so remarkable, what distinguishes the living from the nonliving, is not what organisms are made of but how they are put together and function as wholes.

Even though vitalism is discredited, a germ of the idea is correct. There *is* a nonmaterial "something" inside living organisms, something unique and, literally, vital to their operation. It is not an essence or a force or an atom with a zing. That extra something is a certain type of information or, to use the modern jargon, software.

The tale of the ancient molecule

Inside each and every one of us lies a message. It is inscribed in an ancient code, its beginnings lost in the mists of time. Decrypted, the message contains instructions on how to make a human being. Nobody wrote the message; nobody invented the code. They came into existence spontaneously. Their designer was Mother Nature herself, working only within the scope of her immutable laws and capitalizing on the vagaries of chance. The message isn't written in ink or type, but in atoms, strung together in an elaborately arranged sequence to form DNA, short for deoxyribonucleic acid. It is the most extraordinary molecule on Earth.

Human DNA contains many billions of atoms, linked in the distinctive form of two coils entwined in mutual embrace. This famous double helix is in turn bundled up in a very convoluted shape. Stretch out the DNA in just one cell of your body and it would make a thread two meters long. These are big molecules indeed.

Although DNA is a material structure, it is pregnant with meaning. The arrangement of the atoms along the helical strands of your DNA determines how you look and even, to a certain extent, how you feel and behave. DNA is nothing less than a blueprint—or, more accurately, an algorithm or instruction manual—for building a living, breathing, thinking human being.

We share this magic molecule with almost all other life forms on Earth. From fungi to flies, from bacteria to bears, organisms are sculpted according to their respective DNA instructions. Each individual's DNA differs from others in the same species (with the exception of identical twins), and differs even more from that of other species. But the essential structure—the chemical makeup, the double-helix architecture—is universal.

DNA is incredibly, unimaginably ancient. It almost certainly existed three and a half billion years ago. It makes nonsense of the phrase "as old as the hills": DNA was here long before any surviving hills on Earth. Nobody knows how or where the first DNA molecule formed. Some scientists even speculate that it is an alien invader, a molecule from Mars perhaps, or from a wandering comet. But however the first strand of DNA came to exist, our own DNA is very probably a direct descendant of it. For the crucial quality of DNA, the property that sets it apart from other big organic molecules, is its ability to replicate itself. Put simply, DNA is in the business of making more DNA, generation after generation, instruction manual after instruction manual, cascading down through the ages from microbes to man in an unbroken chain of copying.

Of course, copying as such produces only more of the same. Perfect replication of DNA would lead to a planet knee-deep in identical single-celled organisms. However, no copying process is totally reliable. A photocopier may create stray spots, a noisy telephone line will garble a fax transmission, and a computer glitch can spoil data transferred from hard disk to a floppy. When errors occur in DNA replication, they can manifest themselves as mutations in the organisms that inherit them. Mostly a mutation is damaging, just as a random word change in a Shakespeare sonnet would likely mar its beauty. But occasionally, quite by chance, an error might produce a positive benefit, conferring an advantage on the mutant. If the advantage is life-preserving, enabling the organism to reproduce itself more efficiently, then the miscopied DNA will out-replicate its competitors and come to predominate. Conversely, if the copying error results in a less well-adapted organism, the mutant strain will probably die out after a few generations, eliminating this particular DNA variant.

This simple process of replication, variation, and elimination is the basis of Darwinian evolution. Natural selection—the continual sifting of mutants according to their fitness—acts like a ratchet, locking in the advantageous errors and discarding the bad. Starting with the DNA of some primitive ancestor microbe, bit by bit, error by error, the increasingly lengthy instructions for building more complex organisms came to be constructed.

Some people find the idea of an instruction manual that writes itself simply by accumulating chance errors hard to swallow, so let me go over the argument once more, using a slightly different metaphor. Think of the information in human DNA as the score for a symphony. This is a grand symphony indeed, a mighty orchestral piece with hundreds of musicians playing thousands of notes. By comparison, the DNA of the ancient ancestor microbe is but a simple melody. How does a melody turn into a symphony?

Suppose a scribe is asked to copy the original tune as a musical score. Normally the copying process is faithful, but once in a while a quaver becomes a crotchet, a C becomes a D. A slip of the pen introduces a slight change of tempo or pitch. Occasionally a more serious error leads to a major flaw in the piece, an entire bar omitted or repeated perhaps. Mostly these mistakes will spoil the balance or harmony, so that the score is of no further use: nobody would wish to listen to its musical rendition. But very occasionally the scribe's slip of the pen will add an imaginative new sound, a pleasing feature, a successful addition or alteration, quite by chance. The tune will actually improve, and be approved for the future. Now imagine this process of improvement and elaboration continuing through trillions of copying procedures. Slowly but surely, the tune will acquire new features, develop a richer structure, evolve into a sonata, even a symphony.

The crucial point about this metaphor, and it cannot be stressed too strongly, is that the symphony comes into being without the scribe's ever having the slightest knowledge of, or interest in, music. The scribe might have been deaf from birth and know nothing whatever of melodies. It doesn't matter, because the scribe's job is not to compose the music but to copy it. Where the metaphor fails is in the selection process. There is no cosmic musician scrutinizing the score of life and exercising quality control. There is only nature, red in tooth and claw, applying a simple and brutal rule: if it works, keep it; if it doesn't, kill it. And "works" here is defined by one criterion and one criterion only, which is replication efficiency. If the mistake results in more copies made, then, by definition, without any further considerations, it works. If A out-replicates B, even by the slightest margin, then, generations on, there will be many more A's than B's. If A and B have to compete for space or resources, it's a fair bet that A will soon eliminate B entirely. A survives, B dies.

Darwinism is the central principle around which our understanding of biology is constructed. It offers an economical explanation of how a relatively simple genetic message elaborates itself over the eons to create molecules of DNA complex enough to produce a human being. Once the basic manual, the precursor DNA, existed in the first place, random errors and selection might gradually be able to evolve it. Good genes are kept, bad genes are discarded. Later I shall discuss the adequacy of this austere explanation, but for now I am more concerned with the starting point. Obviously Darwinian evolution can operate only if life of some sort already exists (strictly, it requires not life in its full glory, only replication, variation, and selection). Darwinism can offer absolutely no help in explaining that all-important first step: the origin of life. But if the central principle of life fails to explain the origin of life, we are left with a problem. What other principle or principles might explain how it all began?

To solve this problem, we must seek clues. Where can we look for clues about the origin of life? A good place to begin is to ask where life itself began. If we discover the place where life started, we may be able to guess the physical conditions that accompanied its genesis. Then we can set about studying the chemical processes that occur in such conditions, and build up an understanding of the prebiotic phase bit by bit.

Microbes and the search for Eden

When I was a youngster I was occasionally coerced into attending Sunday school, an ordeal which I hated. The only positive memory I have is of browsing through a picture book describing the Garden of Eden. The image it conjured up was of a well-ordered parkland in which the sun always shone and exotic animals roamed without fear, presumably being entirely vegetarian. It was a nice contrast to life in a dreary London suburb. Unfortunately, the biblical Garden of Eden turned out to be a myth. Still, there must have been a place where Earth's earliest creatures lived, a sort of scientific Eden. Where was it located?

I am writing this section of the book on a showery spring day in the Adelaide hills. The winter rain has turned the countryside green, and everywhere I look a luxuriant canopy of trees towers over a profusion of smaller bushes, shrubs, and grasses. Birds swoop in the sky and flash colorfully between the branches. Hidden among the foliage are snakes, lizards, spiders, and insects. There will also be rabbits, possums, mice, echidnas, and the occasional koala or kangaroo. Even in this arid country, life is conspicuous and exuberant.

The sheer variety of living things has delighted people for thousands of years. But it is only comparatively recently, with the invention of the microscope, that the true diversity of life on Earth has been revealed. For, even as naturalists marveled at the biological richness of a rain forest or a coral reef, a still greater cornucopia lay unseen all around them. This invisible biosphere is the realm of the micro-organisms, single-celled specks of life that inhabit almost every available nook and cranny the planet can provide. Long dismissed as "mere germs," microbes are now known to dominate the tree of life. "You could go out into your back yard," says John Holt of Michigan State University, "and if you really put your mind to it, you could find a thousand new species in not much time."[5] Holt's comment seems exaggerated until you realize that a spoonful of good-quality soil may contain ten *trillion* bacteria representing ten thousand different species! In total, the mass of micro-organisms on Earth could be as great as a hundred trillion tons—more than all the visible life put together.

To be sure, the physical effects caused by micro-organisms are often very visible: through infectious diseases, the fermentation of alcohol, and the degeneration of food, for example. Even so, microbes have been persistently underrated by humans, perhaps because they are so much smaller than we. Stephen Jay Gould believes we should correct this chauvinism by referring to the present era as the Age of Bacteria, so thoroughly do these tiny creatures overwhelm all others in population numbers and variety.[6] By contrast, so-called higher organisms like humans, dogs, and primroses occupy just a few of the peripheral branches of the tree of life.

Size is not the only reason why microbes tend to get overlooked. They aren't easy to culture in the laboratory, and in the wild a lot of them are inert. Also, many different species of bacteria appear superficially identical, and until recently microbiologists tended to lump them together in classification schemes. Now, with the powerful techniques of molecular sequencing, the real genetic differences are revealed. Bacteria that look the same under the microscope may turn out to share fewer genes with each other than they do with humans.

Gould points out that it has always been the Age of Bacteria. Indeed, for most of the duration of life on Earth, it has consisted of nothing but microbes. This sobering fact offers an opportunity, though. Because life began with microbes, we can expect to find important clues about the origin of life by studying living examples. The hope is that some of them will contain relics of their distant past in the form of unusual structures. Vestiges of ancient biochemistry may have been retained as redundant features—the microbial equivalent of the human appendix. It is even possible that living microbes are carrying around within them molecular remnants of a prebiotic world.

By piecing together fragments of information from living microbes, we may be able to work out what the ancestral organism might have been like, and to guess where and how it lived. Unfortunately, you can't tell just by looking what the evolutionary history of micro-organisms might be. They have few anatomical features by which to classify them. No arms or legs, gills or lungs, eyes or ears present themselves for comparison. As I shall explain later, the evidence linking microbes to their ancient ancestors lies largely in their biochemistry—in their genetic makeup and the metabolic pathways they employ. The techniques of modern molecular biology permit this evidence to be teased out. Like scraps of an ancient scroll covered in a half-forgotten text, this trail of molecular evidence, partly obliterated by the ravages of time, offers a seductive glimpse of an evolutionary past stretching back nearly four billion years.

Given that there are so many species of microbes, where should the search for molecular clues be concentrated? Today it is the aerobic and photosynthesizing bacteria that we most notice, but for over two billion years there was little or no free oxygen available on Earth. Yet microbes flourished in a variety of habitats, fermenting alcohol, producing methane, reducing sulfate. Some microbes maintain their ancient lifestyles today, and these are the ones most likely to offer clues to the earliest forms of life. Which suggests an intriguing idea: suppose there survives today an obscure niche, an exotic place, where conditions resemble the asteroid-battered, gas-shrouded, boiling inferno that was the primeval Planet Earth? If we look carefully, we might find relic organisms still living there, microbes that have changed little since the dawn of life.

Is this possible? Could there be such a place?

The answer is yes. And its location is as surprising as it is obscure. Deep beneath the sea, on the dark ocean floor, there are regions where the Earth's crust stretches and tears. Driven by powerful thermal forces deep inside the planet, the rocky strata of the seabed are continually shifting and straining. Here and there, along mid-ocean ridges, the crust is rent to expose molten rock to the icy ocean above. The oozing lava shrinks and cracks as it cools, creating a matrix of fissures and tunnels through which water circulates by convection, dissolving minerals as it goes. At the vents, the Earth spews forth a stream of searing fluid, liberally spiced with chemicals. The brutal encounter of superheated liquid with cold seawater creates chemical and thermal pandemonium.

It seems impossible to imagine that any form of life could inhabit such a harsh environment, more reminiscent of Hades than the Garden of Eden. Yet it does. Astonishingly, these volcanic ocean vents are home to a rich variety of microbes, some of them apparently relics of an ancient biology. Here in the black volcanic depths dwell the closest organisms we know to the first living creatures on Earth. In the coming chapters I shall describe how startling discoveries of submarine and subterranean superbugs are transforming our thinking about the origin of life and the possibility of life on Mars and elsewhere.

But first I need to explain some of the basic principles of biochemistry. Foremost among these are the laws of thermodynamics.

CHAPTER 2

Against the Tide

And, *departing, leave behind us*
Footprints in the sands of time.

H. W. LONGFELLOW[1]

WHEN I WAS A CHILD, my trips to the seaside were rare and valued occasions. Some of my most vivid memories are of beaches. Besides the seaweed and the jellyfish, and the rise and fall of the ocean, I can remember being struck by the sight of strange little holes in the smooth sand left as the tide retreated. These holes were adorned by neat mounds of sand drawn out into slender sausages and folded over and over, like toothpaste squeezed from a tube into a pile. What, I wondered, caused these peculiar formations? I never saw one in the process of appearing, and they would always be washed away again, along with my footprints, by the incoming tide.

I now know that the mounds of sand are made by tiny crabs that burrow under the surface and kick out the detritus, though I am still mystified by how they create the sausage shapes. However, the point is that I was in no doubt, even at a tender age, that some sort of living creature was responsible for them. Of course, there are many patterns in nature not made by biological activity. Indeed, on the very same beach where I saw the mounds there were also rows of firm ridges, formed by the rippling flow of water across the sand. But the toothpaste piles seemed altogether too contrived, too complicated, to be the work of blind inanimate forces. The tidal flow destroyed the little mounds; I did not believe it also created them.

One of the principal ways in which life distinguishes itself from the rest of nature is its remarkable ability to go "against the tide" (in the above example literally) and create order out of chaos. By contrast, inanimate forces tend to produce disorder. There is in fact a very basic law of nature at work here, called the second law of thermodynamics. To understand how life began, we first need to know how it copes with the vagaries of this law.

The degeneration principle

In the last chapter I remarked that living cells are in some respects like tiny machines. All machines need fuel to run. Animals eat food for fuel, whereas plants are solar-powered. An unavoidable byproduct of fuel consumption is heat. This is very familiar from our own bodies: human beings stay warm because of the waste heat from their combustion of food. Heat is also a form of energy, and can drive physical and chemical changes. In the nineteenth century, scientists and engineers were keen to understand the interplay of heat, work, and chemical reactions to help them design more efficient steam engines and other devices. One result of their investigations was the discovery of the laws of thermodynamics. Of these, the second law is the most relevant to the nature of life.

In essence, the second law of thermodynamics forbids the creation of a perfect machine, or *perpetuum mobile*. It acknowledges that all large-scale physical processes are less than 100 percent efficient: there is inevitable waste, or degeneration. Steam engines, for example, do not use all the energy liberated by the coal that is burned; much of the heat from the boiler radiates away uselessly into the environment, and some of the kinetic energy is lost to friction in the moving parts. A good way to characterize this waste is in terms of order and disorder, or useful and useless energy. The motion of the steam locomotive along the track represents ordered or useful energy; the waste heat is disordered or useless energy. Heat is disordered energy because it is the chaotic motion of molecules. It is useless because it is randomly distributed. The second law describes the inevitable and irreversible trend from ordered to disordered forms of energy. Without a supply of fuel, or useful energy—often called "free" energy—the steam locomotive would soon run out of puff.

The second law of thermodynamics is not restricted to engineering. It is a fundamental law of nature; there is no escaping it. The British astronomer Sir Arthur Eddington regarded it as occupying the supreme position among the laws of nature. He once wrote, "if your theory is found to be against the second law of thermodynamics I can give you no hope; there is nothing for it but to collapse in deepest humiliation."[2] It is easy to find everyday examples of the second law at work, cases where order surrenders to chaos. The destruction of sand piles and footprints I have already mentioned. Think also of a melting snowman or a breaking egg. All these processes produce disordered states of matter from relatively ordered ones. The changes are irreversible. You won't see the tide create a footprint or the sunshine make a snowman. And even the king's horses and men were unable to put Humpty Dumpty together again.

Physicists measure the loss of useful energy in terms of a quantity termed entropy, which roughly speaking corresponds to the degree of chaos present in the system. When a physical process occurs, such as a piston-and-cylinder cycle in a steam engine, it is possible to compute how much entropy is produced as a result. Armed with the concept of entropy, we can state the second law as follows: In a closed system the total entropy cannot go down. Nor will it go on rising without limit. There will be a state of maximum entropy or maximum disorder, which is referred to as thermodynamic equilibrium; once the system has reached that state it is stuck there.

To make these principles clear, let me illustrate them with a simple example concerning the direction of heat flow. If a hot body is put in contact with a cold body, heat passes from hot to cold. Eventually the two bodies reach thermodynamic equilibrium—i.e., a uniform temperature. The heat flow then ceases. Why is this a transition from order to disorder? The uneven distribution of heat at the start can be regarded as a relatively more ordered, hence lower-entropy, state than the final one, because in the final state the heat energy is distributed chaotically among the maximum number of molecules. In this example, the second law demands that heat flow from hot to cold, never the other way.

When the laws of thermodynamics are applied to living organisms, there seems to be a problem. One of the basic properties of life is its high degree of order, so, when an organism develops or reproduces, the order increases. This is the opposite of the second law's bidding. The growth of an embryo, the formation of a DNA molecule, the appearance of a new species, and the increasing elaboration of the biosphere as a whole are all examples of an increase of order and a decrease of entropy.

Some eminent scientists have been deeply mystified by this contradiction. The German physicist Hermann von Helmholtz, himself one of the founders of the science of thermodynamics, was one of the first to suggest that life somehow circumvents the second law.[3] Eddington likewise perceived a clash between Darwinian evolution and thermodynamics, and suggested either that the former be abandoned or that an "anti-evolution principle" be set alongside it.[4] Even Schrödinger had his doubts. In his book *What Is Life?* he examined the relationship between order and disorder in conventional thermodynamics and contrasted it with life's hereditary principle of more order from order. Observing that an organism avoids decay and maintains order by "drinking orderliness" from its environment, he surmised that the second law of thermodynamics may not apply to living matter. "We must be prepared to find a new type of physical law prevailing on it," he wrote.[5]

So is there a problem with the second law of thermodynamics when it comes to biological organisms? No, there isn't. There is no conflict between life and the laws of thermodynamics. To see why not, consider first the case of the humble refrigerator, which is designed precisely to transfer heat energy from a cold place (the inside of the refrigerator) to a warmer place (the kitchen). I stated above that heat is required to flow always from hot to cold, but there is an important condition. The second law stated in this form applies only to *closed* systems. A refrigerator is not a closed system. To force heat to flow "the wrong way," a refrigerator must do some work. This requires a motor and some fuel to drive it. The motor expends energy, irreversibly, and this raises the entropy of the kitchen. When the sums are done, you find that, sure enough, the entropy inside the refrigerator goes down, but the entropy of the kitchen goes up by an even greater amount. (The motor of the refrigerator gets hot when it is running.) What is gained on the swings is more than lost on the roundabouts. So, on balance, running a refrigerator raises the entropy of the universe a bit. The same is true of all processes, including life, that seem to create order out of chaos. They may make order in one place, but they will inevitably make disorder somewhere else to pay for it.

It is not hard to trace where the disorder appears in biological systems. To grow, an organism needs energy or fuel. Food contains useful energy, some of which is dissipated as waste heat during respiration. It is this heat that keeps us warm, and to that extent it is useful, but inevitably some of it flows away into the air around us and is wasted. Thus the burning of foodstuffs in our bodies generates entropy—more than enough to pay for the additional order represented by the production of new cells. The story with plants is similar. Plants grow by capturing solar energy; the transfer of light from the hot Sun to the cool Earth involves a rise in entropy, which more than offsets the increase of order due to the manufacture of new cells.

The second law can also be applied to biological evolution. The appearance of a new species marks an increase in order, but Darwin's theory identifies the price that is paid to achieve this. To evolve a new species requires many mutations, the vast majority of which are harmful and get eliminated by the sieve of natural selection. For every successful surviving mutant, there are thousands of unsuccessful dead ones. The carnage of natural selection amounts to a huge increase in entropy, which more than compensates for the gain represented by the successful mutant.[6]

The upshot, then, is that biological organisms comply fully with the second law of thermodynamics. As long as the environment can provide a supply of free energy, biological systems can go on merrily reducing entropy and increasing order in their local neighborhood, while at the same time contributing to the remorseless rise in the entropy of the universe as a whole. This straightforward resolution of the thermodynamic problem of life was already identified long ago by another of the founders of the theory of thermodynamics, the Austrian physicist Ludwig Boltzmann: "Thus, the general struggle for life is neither a fight for basic material . . . nor for energy . . . but for entropy becoming available by the transition from the hot sun to the cold earth."[7]

We must be careful, however, not to fall into a trap here. That life is consistent with the second law of thermodynamics does not mean that the second law *explains* life. It certainly doesn't. Unfortunately, many scientists who should know better have succumbed to this fallacy. We still have to demonstrate *how* the exchange of entropy with the environment brings about the very specific sort of order represented by biological organisms. Merely specifying a source of useful energy does not of itself offer an explanation for how the ordering process happens. To do that, one needs to identify the exact mechanisms that will couple the reservoir of available energy to biologically relevant processes. To overlook this part of the story is rather like proclaiming that the function of refrigerators is explained once we have found an electric socket.

Because it corresponds to equilibrium, a state of maximum entropy is stable. Conversely, a state of thermodynamic disequilibrium is unstable; natural processes want to drive the entropy up to a maximum. However, in practice there may be barriers preventing the second law from having its way. For example, a mixture of gasoline fumes and air is not a maximum-entropy state. The two gases would like to react to form more stable substances and liberate heat, thus raising the entropy. Under normal conditions, this reaction is stymied: a chemical barrier prevents it from happening spontaneously. It requires a spark to trigger the reaction. States that have a fragile stability of this sort are termed metastable. A mixture of gasoline fumes and air is one example of a metastable state. Another is a pencil stood on its flat end. It needs a little shove to make it topple over—in contrast to a pencil perched on its tip, which is completely unstable.

The concept of metastability is absolutely crucial to the success of life. Living organisms get their useful energy from chemical reactions, but they could not do this if inorganic processes had short-circuited the process and squandered the energy first. So life is always on the lookout for metastable sources of free energy to exploit. Animals derive their energy by burning organic material, making use of the same basic metastability as the gasoline–air mix. As we shall see, some microbes extract energy by seeking out chemical pathways that even chemists wouldn't think of.

To tap into metastable sources, organisms have to overcome the activation barriers that frustrate the inorganic release of the energy. They do this by clever strategies, such as the use of enzymes to catalyze reactions that would otherwise proceed extremely slowly. Another of their tricks is to deploy energized molecules to add the equivalent of the spark that ignites the gasoline. Because chemical reactions take place at exceedingly different rates under different circumstances, organisms can control the release of energy, delivering small doses when and where needed. This fact makes chemistry the ideal basis for biology, but in principle life could function using any metastable energy source. Science-fiction writers have speculated about life based on ionized plasma or nuclear processes. Though this may be theoretically possible, the sheer variety and versatility of chemical reactions must make chemical life by far the best bet.

Where does biological information come from?

Modern warfare depends heavily on reliable communications. Telephone lines and radio links have long played a crucial role in military command and control. Yet both these communication channels are subject to signal interference, as anyone who has tried to relay instructions over an out-of-range mobile phone is well aware. During the Second World War, the United States Army commissioned a study of the principles of communication by Claude Shannon, a researcher at the Bell Telephone Laboratories. The results of his analysis were published in 1949 under the title *The Mathematical Theory of Communication*, and the book soon became a classic.[8]

Shannon's theory hinges on a direct link between information and entropy. Imagine talking to a friend over a hissing telephone line. It goes without saying that the background noise never adds anything to the conversation, though it may prevent you from receiving some information. Shannon's great insight was to spot that noise is a form of disorder, or entropy. By contrast, a signal represents order: compare the carefully arranged dots and dashes of Morse code with the crackle of radio static. In Shannon's theory, information is treated as the opposite of entropy; for that reason, information is sometimes referred to as negative entropy. When information is lost in a noisy communication channel, the entropy rises. This is therefore another example of the ubiquitous second law of thermodynamics. So signal degradation can be regarded in two equivalent ways: as noise invading the channel, or as information leaking away. This new slant on entropy can be applied quite generally to physical systems. The second law can be thought of either as a rise in entropy, or as a decline in the information content of the system.

Shannon's ideas have obvious application to biological organisms, because information is one of their defining properties. DNA stores the information needed to construct and operate the organism. One aspect of the mystery of biological order can therefore be expressed by the question: where does biological information come from? Communication theory—or information theory, as it is known today—says that noise destroys information, and that the reverse process, the creation of information by noise, would seem to us a miracle. A message emerging on its own from radio static would be as surprising as the tide making footprints on the beach. We are back with the same old problem: the second law of thermodynamics insists that information can no more spring into being spontaneously than heat can flow from cold to hot.

The solution to the problem may once again be found in the fact that an organism is not a closed system. The information content of a living cell can rise if the information in its surroundings falls. Another way of expressing this is that information flows from the environment into the organism. This is essentially what Schrödinger meant when he said that an organism makes a living by "drinking orderliness." Life avoids decay via the second law of thermodynamics by importing information, or negative entropy, from its surroundings. The source of biological information, then, is the organism's environment.

Both metabolism and reproduction are driven by information flow from environment to biosystem. Food contains ordered or useful energy, rich in information; think of the complex organic molecules as analogous to little bits of Morse code. Body heat is wasted energy, information-poor, like a telephone line that just hisses. Thus the second law exacts its toll, but the organism grows nonetheless by concentrating information within itself and exporting the entropy. In the case of reproduction, the information content of DNA changes much more slowly—over many generations—as a result of random mutations. Mutations are the biological equivalent of noise in a telephone line. The "signal" is the freshly minted DNA. Successful mutations are those that are better adapted to their environment, and it is therefore the environment that provides—or, more accurately, selects—the information that ends up in the DNA. So the environment feeds the information into the genetic message via natural selection.[9]

Viewing the struggle for existence in terms of the ebb and flow of information raises a curious question. Are mutations good or bad news? If genome replication were completely faithful, life could never adapt to changing circumstances, and extinction would inevitably follow. On the other hand, too many copying errors and the genetic message would get diluted and eventually lost. To succeed, a

species needs to strike a balance between too many and two few mutations.

We can see this compromise being acted out in our own lives.
When I was seven years old, an elderly aunt of mine died of tuberculosis. This was the first I had ever heard of the once-feared consumption, or TB, and it was also to be the last for quite some time.
Even as early as the 1950s, death from this age-old scourge was becoming rare in Britain, and the rate was to decline rapidly over the
coming decade to almost negligible proportions. The discovery of
the antibiotic streptomycin in 1943, and the subsequent use of the
BCG vaccine, effectively eliminated TB as a public-health issue.
Until now. Suddenly, tuberculosis is back in the news as the latest
drug-resistant killer. Along with new strains of salmonella, gonor-
rhea, and pneumonia, tuberculosis threatens to become a major
health hazard once again. What is happening?

Part of the answer lies in the capacity of bacteria to multiply extremely fast, combined with their high rate of mutation. This almost
guarantees that they will outmaneuver whatever drugs the medical
profession throws at them. As fast as researchers come up with a
new antibiotic, the ever-changing pathogens jump one step ahead.

The tussle between doctors and bacteria is a good example of
Darwinian evolution at work. Though the situation with infectious
diseases is complicated by various medical factors, a simple principle
can be discerned in the underlying replication process. As I have explained, mistakes in information transmission are like noise, or entropy, in a communication channel. Noise causes information to
leak away—in this case genetic information. This degradation of
the genetic message is countered by natural selection, which serves
as a source of information. If the environment cannot put back into
the genome via natural selection as much information as leaks out,
the errors will eventually accumulate to the point where they mess
up the replication process itself, and reproduction will cease. This
disastrous state of affairs, which is just another example of the second law of thermodynamics at work, has been dubbed "the error catastrophe" by the German biochemist Manfred Eigen.

The error catastrophe can be quantified by asking how many bits of information there are in an organism, and how much of it can leak away before that particular lineage succumbs. Eigen has demonstrated that, the greater the number of genes the organism possesses, the lower the error rate must be to avoid the error catastrophe, in simple proportion. In other words, sloppy copying kills complex organisms. A higher organism has about a hundred thousand genes, capable of storing about a hundred million bits of information, each of which may be subject to copying errors. As a rough estimate, if the error rate is less than one in a hundred million per replication, the error catastrophe will be avoided. By contrast, bacteria, which have far fewer genes, can get away with much higher error rates. Nature seems to know Eigen's rule. Cells like ours manage to cut back their error rate to about one in a billion, whereas for bacteria it is much higher—about one in a million. Hence the problems with drug-resistant mutations. For viruses, which have even fewer genes than bacteria, the mutation rate can be higher still. The optimum error rate for a species will normally be just below the error catastrophe, for this provides a compromise between stability and flexibility.

The error catastrophe is crucially important for the problem of biogenesis. In modern organisms, sophisticated proofreading and error-correction mechanisms are employed to keep the error rate down. Cells can call upon a suite of enzymes, evolved over billions of years, to finesse the copying process. No such enzymes would have been available to the first organisms. Their replication must have been extremely error-prone. According to Eigen's rule, this means that the genomes of the first organisms (or the prebiotic replicators) must have been very short in length if they were to evade the error catastrophe. But here we hit a paradox. If a genome is too short, it can't store enough information to build the copying machinery itself. Eigen believes that even the simplest replication equipment requires much more information than could ever have been accommodated in a primitive nucleic-acid sequence.[10] To reach the sort of length needed to code for the necessary copying enzymes, the genome risks falling foul of the very error catastrophe it is trying to combat. To put it simply: complex genomes demand reliable copying, and reliable copying requires complex genomes. So which came first? Such chicken-and-egg paradoxes are typical of the problems of biogenesis, as we shall see in chapter 5.

So far, I have been somewhat cavalier in the use of the term "information." Computer scientists draw a distinction between syntax and semantics. Syntactic information is simply raw data, perhaps arranged according to rules of grammar, whereas semantic information has some sort of context or meaning. Information *per se* doesn't have to mean anything. Snowflakes contain syntactic information in the specific arrangement of their hexagonal shapes, but these patterns have no semantic content, no meaning for anything beyond the structure itself. By contrast, the distinctive feature of biological information is that it is replete with meaning. DNA stores the instructions needed to build a functioning organism; it is a blueprint or an algorithm for a specified, predetermined product. Snowflakes don't code for, or symbolize, anything, whereas genes most definitely do. To explain life fully, it is not enough simply to identify a source of free energy, or negative entropy, to provide biological information. We also have to understand how *semantic* information comes into being. It is the quality, not the mere existence, of information that is the real mystery here. All that stuff about conflict with the second law of thermodynamics was mostly a red herring.

The source of semantic information can only be the environment of the organism, but this begs the question of how the information got into the environment in the first place. It is surely not waiting, like fragments of a pre-existing blueprint, for nature to assemble it. The environment is not an intelligent designer. So what do we know about the information content of the environment itself? Indeed, what is meant here by "the environment"? The organism's habitat? The biosphere? The solar system? In the end, the environment is the entire universe. Follow the chain of causation and the question becomes one of cosmology. We are then confronted by the ultimate question: where did the information content of the universe come from?

The entropy gap: gravity as the fountainhead of order

Darwin once chided those who would speculate about the origin of
life with the retort that one might as well speculate about the origin
of matter. Today, physicists and cosmologists think they know how
matter originated, and it turns out to be extremely revealing to compare the process with biogenesis. The observable universe contains
about 10^{50} tons of matter, and the problem of where it came from
plagued cosmology for many years. Early critics of the big-bang theory rightly objected to the assumption that all this matter just
popped into existence at the beginning of time for no apparent reason. The idea that the universe originated with the necessary matter
already there, *ab initio*, struck many as totally unscientific.

A way forward lay at hand, however. Physicists long ago discovered that particles of matter can be created if enough energy is concentrated, a process that can actually be demonstrated in the
laboratory using large accelerator machines. Unfortunately, this
didn't quite solve the cosmological problem, because it begged the
question of where the energy needed to make the cosmological material came from in the first place. The assumption that the energy
of the universe was simply "given"—i.e., it was there at the outset—
was hardly an improvement on the assumption that matter was
there at the outset. There thus remained an element of miracle, of
something-for-nothing, in the big-bang theory.

In the 1980s, the puzzle of the source of cosmic energy was solved.
It was discovered that the total energy of the universe might actually
be zero, and it was therefore really a case of nothing-for-nothing. The
reason the universe can have zero energy and still contain 10^{50} tons
of matter is that its gravitational field has negative energy—a peculiar
concept related to what I have to say below. The sums show that the
two contributions can exactly cancel to leave zero. A convincing
mechanism was found to explain how positive energy was channeled
into matter, and an equal quantity of negative energy went into the
gravitational field. So, in effect, all the cosmic matter was actually cre-

ated for free! Once cosmologists realized this, it became credible to suppose that at the beginning of the universe space was empty; all the matter appeared later (though still pretty quickly), as a result of a natural physical process. The new theory was regarded as superior and more scientific, because it removed the need to postulate the supernatural input of matter at the beginning of time.

Turning now to the problem of biogenesis, we encounter an odd reversal of sentiment. We now need to explain, not the origin of material stuff, but the origin of information. Whereas it is good science to seek a physical process to generate matter, it is regarded as unscientific in the extreme to entertain a process that generates information. Information is not something that is supposed to come for free (like cosmic matter), but something you have to work for. This is really just the second law of thermodynamics revisited, because the spontaneous appearance of information in the universe would be equivalent to a reduction of the entropy of the universe— a violation of the second law, a miracle. Now, the fact that the universe contains information is undeniable (because it is not in thermodynamic equilibrium). If information can't get made, it must have been there at the beginning, as part of the initial input. The conclusion we are led to is that the universe came stocked with information, or negative entropy, from the word "go."

What do astronomical observations say about the information content of the early universe? Here we make a very curious discovery. One of the most compelling pieces of evidence for the big-bang theory is the existence of a universal background of heat radiation, which seems to be a sort of afterglow of the universe's fiery birth. This radiation has traveled across space more or less undisturbed since shortly after the big bang. It therefore provides a snapshot of what the universe was like near the beginning. Satellite measurements have determined that the spectrum of the cosmic heat radiation corresponds precisely to a state of thermodynamic equilibrium. But thermodynamic equilibrium is a state of maximum entropy that, via the Shannon connection, implies *minimum* information. In fact, it suffices to give just one bit of information (the temperature) to characterize completely a state of thermodynamic equilibrium, so, if the cosmic background heat radiation is anything to go by, the universe started out with almost no information content at all.

We seem to be faced with a disturbing contradiction. The second law forbids the total information content of the universe from going up as it evolves, yet, from what we can tell about the early universe, it contained very little information. So where has the information present in the universe today come from? Another way of expressing the problem is in terms of entropy. If the universe started out close to thermodynamic equilibrium, or maximum entropy, how has it reached its present state of disequilibrium, given that the second law forbids the total entropy to go down?

The answer to this cosmic conundrum is now well known: it comes from a careful study of gravitation. To see how gravitation makes a difference to thermodynamics, think of a flask of gas at a uniform temperature. If the gas is left undisturbed it will do nothing; that is, it will remain in equilibrium. But suppose the mass of gas is so great (as large as an interstellar cloud, say) that gravitation becomes important. Then it is no longer true that nothing happens. The system is now unstable. The gas will start to contract, and clumps of denser material will accumulate here and there. At the centers of the clumps the contraction will make the gas hot. Temperature gradients will form and heat will flow. In a real interstellar cloud, stars form. The flow of heat radiation from one such star—the Sun—is the source of free energy, or negative entropy, that drives all surface life on Earth through photosynthesis. So, under the action of gravitation, a gas that is supposed to be in thermodynamic equilibrium at a uniform temperature and maximum entropy, nevertheless undergoes further changes, causing heat to flow and the entropy to rise further. Thus gravitationally induced instability is a source of information.

Evidently gravitation changes the rules of the game in a profound way. A system in which gravitation makes itself felt cannot be considered to be in a state of true thermodynamic equilibrium, or maximum entropy, just because it is at a uniform temperature and density. Appearances deceive us. A uniform cloud of gas still has a lot of free energy to give up via gravitational processes. Even at a uniform temperature, the gas is in a *low*-entropy state. When it comes to cosmology, gravitation is the all-dominant force, so we cannot ignore its thermodynamic effects. This means we cannot conclude from the existence of a uniform background of heat radiation that the early universe was in fact in a state of overall thermodynamic equilibrium, gravitation included.

Just as life seems to go "the wrong way" thermodynamically, so too does gravitation go "the wrong way."[11] A smooth gas grows into something clumpy and complex. Order appears spontaneously. In informational terms, this seems all back to front. A uniform gas, by its very simplicity, can be described with very little information, whereas a star cluster or a galaxy requires a lot of information to describe it. In some as yet ill-understood way, a huge amount of information evidently lies secreted in the smooth gravitational field of a featureless, uniform gas. As the system evolves, the gas comes out of equilibrium, and information flows from the gravitational field to the matter. Part of this information ends up in the genomes of organisms, as biological information.

Looking at the universe as a whole, the initially smooth distribution of gas coughed out at the big bang slowly turned into splodges of hotter and cooler gas, and eventually arranged itself into shining proto-galaxies surrounded by empty space. The proto-galaxies in turn formed glowing stars. The expansion of the universe assisted the escalating thermal contrast: as the universe expanded, its background temperature dropped, and the hot stars were then able to radiate more vigorously into the cold space. The upshot of these gravitational processes was that an entropy gap opened up in the universe, a gap between the actual entropy and the maximum possible entropy. The flow of starlight is one process that is attempting to close the gap, but in fact all sources of free energy, including the chemical and thermal energy inside the Earth, can be attributed to that gap. Thus all life feeds off the entropy gap that gravitation has created. The ultimate source of biological information and order is gravitation.

Tracing the source of information back to gravitation and the smooth state of the universe just after the big bang still leaves us with the problem of semantics. How has *meaningful* information emerged in the universe? This mystery is closely related to the origin of complexity, another defining factor of life. Scientists are divided over whether complexity behaves like matter or information—that is, whether or not the overall complexity of the universe stays the same. Some researchers are convinced that there are laws of complexity. If such laws exist, they may describe how a simple state can evolve naturally into a more complex one, perhaps even one containing semantic information. This process is often called self-complexification or self-organization, and I shall have more to say about it in the coming chapters. Other scientists argue that complexity cannot be conjured out of midair; a complex system can only be created by another system at least as complex. But gravitational complexity gives pause for thought, because it does indeed emerge naturally from a simple initial state.

As such a weak force, it is hard to see how gravitation could play a direct role in biochemical processes. However, suggestions have been made along those lines. Roger Penrose, an Oxford mathematician and a world expert on gravitation theory, has speculated that gravity may affect biomolecules through quantum processes.[12] Mathematical physicist Lee Smolin has also compared the subjects of life and gravitation in his recent book *The Life of the Cosmos*. He develops an analogy between the behavior of ecosystems and spiral galaxies. Drawing inspiration from computer models of self-organization, Smolin finds close parallels in the processes of feedback and pattern formation in star clusters and biology. He believes that life is part of a "nested hierarchy of self-organized systems that begins with our local ecologies and extends upwards at least to the galaxy."[13]

If these ideas of Penrose and Smolin are right—and it has to be said that they are very speculative—they may reveal a connection between the thermodynamically "wrong-way" qualities that characterize both gravitational and biological systems. It would follow that the explanation of the origin of life is deeply linked to the origin of the universe itself.

In this same speculative vein, I should like to offer some ideas of my own. The concept of information crops up in many different scientific contexts—not just biology and thermodynamics, but in computation and in other branches of physics too. In quantum mechanics, for example, the wavelike aspects of matter are described by a mathematical object known as the wave function, which represents everything that is known about the system being described; i.e., it represents the *information content* of the state. I shall have more to say on this topic in chapter 10. Here I merely wish to remark that the distinctive feature of the wave function is its so-called nonlocality—it is spread out across space and describes mysterious linkages between widely separated particles, linkages that Einstein dubbed "spooky action at a distance." In other words, the wave function, and its information content, is a global entity, not a local quantity like momentum, energy, or electric charge.[14]

In relativity theory, information pops up again, but in a very different, and very curious, context. It is often said that the theory of relativity forbids anything to travel faster than light. That is not true. It does permit particles to travel faster than light (such hypothetical particles are called tachyons). What is forbidden by the theory is the transmission of *information* faster than light. The problem here is that, if A can signal B at superluminal speed, it is easy to devise a setup that can send signals into the past and thereby create classic causality paradoxes.[15] These paradoxes do not result from the possibility of superluminal propagation as such: faster-than-light *noise* is no threat to causality, because it is devoid of information. But faster-than-light *signals* (i.e., information) are deeply paradoxical. For example, imagine that the radio-control device that opens my garage door were able to transmit its signal into the past by, say, one day. I could then place the device on a radio-activated bomb, programmed to explode if it receives a signal from the future. What will happen if I press the button tomorrow? The bomb should explode today, wrecking the device and preventing me from activating it tomorrow. But if I don't activate it tomorrow, the bomb will not explode. Paradoxes of this sort are very familiar to devotees of science fiction. Now, in principle, the trigger for the bomb doesn't have to be a complicated radio signal; it need only be a single quantum particle from the transmitter, as long as the system is set up appropriately to respond to it. In other words, if the system is constructed in such a way that the particle concerned is a signal to explode the bomb, we encounter a paradox. But the particle on its own is undistinguished—a particle is a particle. It becomes the trigger for a bomb, and a paradox, if it conveys *information* from the transmitter to the receiver. That is, the *context* in which the particle travels backwards in time produces the problem. And context is a global concept. The particle cannot on its own betray whether it conveys information or not; no quality that attaches to it locally (as would, for example, electric charge) says, "I possess information."

Thus, both quantum mechanics and relativity suggest that information is a global rather than a local physical quantity. You cannot simply inspect a location in space and detect information. What you see—a particle, for example—becomes information only in an appropriate global context. Yet whether or not the particle does represent information is not a trivial or purely semantic matter. It may have dramatic *physical* consequences, as the bomb example graphically demonstrates.

How does all this relate to the origin of life? It suggests that we will not be able to trace the origin of biological information to the operation of *local physical forces and laws*. In particular, the oft-repeated claim that life is written into the laws of physics cannot be true if those laws are restricted to the normal sort, which describe localized action and proximate forces. We must seek the origin of biological information in some sort of global context. That may turn out to be simply the environment in which biogenesis occurs. On the other hand, it may involve some nonlocal type of physical law, as yet unrecognized by science, that explicitly entangles the dynamics of information with the dynamics of matter.

CHAPTER 3

Out of the Slime

> You expressed quite correctly my views where you
> said that I had intentionally left the question of the
> Origin of Life uncanvassed as being altogether ultra
> vires in the present state of our knowledge.
>
> CHARLES DARWIN[1]

EARL MOUNTBATTEN OF BURMA, late cousin of Queen Elizabeth II, was fond of claiming that he could trace his royal lineage back to a time before the Norman Conquest of 1066. It was an impressive boast. Such a pedigree certainly puts us commoners in our place. Or does it?

A thousand years of history is about forty generations. Each of us had two parents, four grandparents, and eight great-grandparents. For every generation one goes back, the number of ancestors doubles. Using this rule, it seems that forty generations ago I would have had 2^{40} or about a trillion ancestors. That is much more than all the people on Earth who have ever lived, so something must be wrong with the arithmetic.

The mistake is to assume that human ancestry spreads out forever into the past, as family trees suggest. In reality, at some point as you trace a family tree back in time, the lines start to cross and recross. Genes, and royal blood, diffuse across the planet, making us all distant cousins. I too have royal blood in my veins; it's just that, unlike Lord Mountbatten, I don't have the necessary documentation to prove it.

Further thought about family trees leads to a still stranger conclusion. Not only do they fail to spread out forever into the past, they must at some point start to converge. A hundred thousand years ago, there were but a handful of Homo sapiens on the planet, from whom all people alive today, without exception, have descended. By extrapolation, this convergence will terminate on a single hominid ancestor. (In the female line, this ancestral person is popularly referred to as African Eve, because it seems likely that she lived in Africa.) What is good for humans is also good for other species. We share almost all our genes with chimpanzees, for example. A few million years before African Eve walked the savanna, a common ancestor of all apes and humans dwelt somewhere in the African forest. And so on back through time. The deeper one delves into the past, the more interrelated will be the species that are now quite distinct. Half a billion years ago, I had a fish as an ancestor. Two billion years ago, all my ancestors were microbes.

Similar reasoning applies to all organisms, including the bush outside my study, the bird pecking at the window, and the mushrooms on the lawn. If we could follow their family trees far enough back in time, their separate branches would eventually tangle and join. We can envisage a family tree of everything alive today, a sort of supertree of life. Ultimately the branches of this supertree must also converge, not just a little, but completely—converge until they narrow down on to a central trunk. This ancient stem represents a single primitive organism, the common ancestor of all terrestrial life, a microbial Adam whose destiny was to populate the planet with a myriad progeny. But how did this tiny organism, this begetter of a billion species, come to exist? Where did it live, and when? And what came before it?

The tree of life

In the spring and summer of 1837, fresh from his voyage on HMS Beagle, Charles Darwin began the grand synthesis of his research that was to become his celebrated theory of evolution. In mid-July, Darwin's thoughts were still scattered, his mood one of confused groping. In a notebook, amid many tentative doodles and frantic jottings, he made a simple sketch that was to capture at a stroke the conceptual sweep of the theory slowly forming in his mind. The drawing was of an "irregularly branched" tree, intended to convey the genealogical history of plants and animals: a tree of life.[2] As a metaphor it was brilliant, conveying the essential notion that life originated in the dim and distant past with a unique, spontaneous event. From this single common ancestor—the trunk of the tree— life diversified over time by successive branchings, with new species splitting away from old. The ends of the branches represent extinctions, like the dinosaurs and the dodo.

The existence of a solitary trunk was a guess. Darwin disliked what he called the "excessively complicated" notion of life constantly emerging, creating a jumbled forest of life in place of a lone tree. Today, biologists insist that Darwin's guess was basically correct: life on Earth has descended from a single common ancestor.

What makes them so sure? There are several excellent reasons to believe in a universal ancestor. For a start, every known organism shares a common physical and chemical system. The metabolic pathways of the cell—how it grows, which molecules do what and when, how energy gets stored and liberated, where proteins get made and what they do—are basically the same throughout. The way in which a cell records genetic information and reproduces it is also common to all life. Perhaps the most convincing evidence for a common origin is that genetic instructions are implemented using a universal code (see chapter 4). It is too much to believe that all these complex and highly specific features arose independently many times. More likely, they reflect properties already present in a universal ancestor cell, and inherited by its descendants.

Evidence for a common ancestor also comes from the curious matter of molecular handedness—or chirality, as it is known technically. Most organic molecules are not symmetric: their mirror images look different, just as a left hand differs from a right hand (they

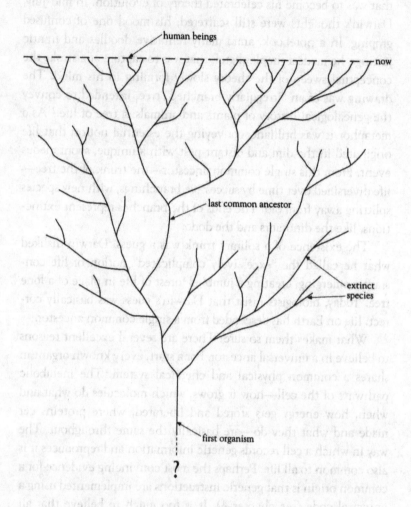

Figure 3.1. The tree of life, greatly simplified. The trunk of the tree represents the first living thing. The present day corresponds to the top branches of the tree, among which human beings are found. The universal ancestor of extant life is located at the last fork in the tree that connects to all the topmost branches. Branches below this fork correspond to organisms that have left no surviving descendants. The diagram as drawn greatly exaggerates the number of extant versus extinct species.

have "opposite chirality"). For example, DNA is coiled like a right-handed helix; its mirror image is a left-handed helix. However, the forces that hold molecules together make no distinction between left and right. No law of nature forbids a left-handed DNA molecule, yet nobody has ever found one. The same chirality, whether left or right, is common to all living things. This suggests that all life descended from a single ancestor cell that contained molecules having the particular chiralities found today.

It is important not to confuse the last common ancestor with the first living thing. To make this point clear, figure 3.1 shows, very schematically, the tree of life as it appears today. Start anywhere along a branch and you can trace a route back to the central trunk. Notice that most of the lower branches, representing life long ago, have terminated. In fact, more than 99 percent of all species that have ever lived are now extinct. If you start at the top of the tree (corresponding to the present day) and follow the branches down to a latest common origin, this point may well lie, not at the base of the central trunk at all, but above some of the lower branches. These lower branches represent extinct species *all* of whose descendants are also extinct. They are quite literally dead-ends on the tree of life.

Most of the dead-ends were undoubtedly creatures resembling extant life in their basic biochemistry. It is conceivable, however, that some were cells using exotic processes not found in any species alive today. For example, there may have been microbes that employed a different genetic code. These exotics might have found themselves in stiff competition with "our" sort of life, and been driven to extinction because they were less well adapted. It is also possible that they may not have died out completely. Perhaps someday biologists will stumble across weird microbes in an unusual niche somewhere on Earth, or on Mars, that turn out to be surviving descendants of one of these lower branches in the tree of life. This microbial Lost World would give scientists a wonderful chance to study otherwise obsolete metabolic or genetic processes.

Intriguingly, our own metabolism may contain harmless remnants of an alternative biochemical system long ago discarded by our ancestors, but fatally retained by now extinct organisms. If so, we will have within our very bodies the faint memory of an alternative life form that became extinct billions of years ago. This idea is not as speculative as it may seem. Many cells (including our own) contain little subunits known as mitochondria. These structures are believed to be the vestiges of once-independent microbes that invaded host cells and took up permanent residence. It is a process called symbiosis.

To see how microbial symbiosis might come about, imagine a typical battle among bacteria. Microbes will attack and eat each other with the same ruthlessness as lions and sharks in their struggle for survival. At the bacterial level, however, the process of ingestion and the process of infection are really the same—A ends up inside B. If B wins and A dies, we called it eating; if A wins and B dies, we call it infection. However, it may happen that A and B reach a stalemate and come to an accommodation: both A and B survive in a symbiotic relationship. There are many examples of symbiosis in nature, such as useful parasites. We need look no farther than our own alimentary tract, which swarms with bacteria that assist us in digesting food and make a healthy living for themselves as a result. We couldn't get along without such bacteria—still less without mitochondria, which act as crucial energizing units for cells.

The theory that mitochondria were once free-living organisms is a century old, but it was championed most persuasively in the late 1960s by Lynn Margulis. According to the theory, mitochondria would initially have employed their own metabolic and reproductive processes, in peaceful coexistence with their hosts. Over time, however, evolution has stripped them of most of their original autonomy, and their activities have been subjugated to the agenda of the host cell. But mitochondria still retain some of their original genetic material—a faint memory of their erstwhile independence.

Since Margulis published her theory, the evidence has grown in support of it. Now it seems that not only mitochondria but other structures within cells, such as microtubules, whiplike flagella, and peroxisomes—blobs within membranes that protect cells against oxygen poisoning—might also be vestiges of bacterial invaders. In green plants, chloroplasts, which carry out the vital function of photosynthesis, are probably descended from cyanobacteria. So some branches of the tree of life may end up fusing with others rather than terminating in dead-ends.

The three domains of life

When I was at school, we were taught that living things were divided into two great kingdoms: plants and animals. Some single-celled creatures, like amoebae, were treated as rudimentary animals, whereas algae were regarded as simple plants. Questions about bacteria were discouraged. Regrettably, we were being misled. In 1937, a better classification scheme was introduced that divided life into two rather different domains, called prokaryotes and eukaryotes. Prokaryotes are small, relatively primitive single-celled organisms lacking cell nuclei and other complicated structures. They include bacteria. Eukaryotes make up the rest. They consist of larger and more complex single-celled organisms such as amoebae, plus all multicellular organisms, which can be thought of as colonies of eukaryotic cells. Although the great proliferation of multicellular life didn't start until about six hundred million years ago, the eukaryotes paved the way for it much earlier.

The tree shown in figure 3.1 is purely schematic. Fortunately, a more quantitative version can be worked out in which the degree of genetic differences between the branches is displayed. Because a cell is subject to copying errors when it reproduces, initially identical cells can drift apart, genetically speaking, as mutations accumulate over time. If enough mutations occur, a new species arises. As a general rule, the greater the number of changes between two sets of genes, the farther apart are the species situated on the tree of life. For example, your genes are very similar indeed to mine, less similar to those of an ape, and still less to those of a tortoise, or a pea. The differences in genetic makeup can be measured rather precisely using gene- and protein-sequencing techniques, and the relative positions on the tree of life computed.

The procedure can be compared to the study of the evolution of languages. When the Vikings first settled Iceland, they spoke a common tongue with their Scandinavian forebears. Over time, however, lack of contact between the settlers and mainland Europe ensured that Icelandic diverged from the mother tongue to the extent that it is now recognized as a separate language in its own right. But go back five hundred years and the differences would not have been so great. The degree of divergence of two languages thus gives a measure of how long the two nations have been developing separately.

About thirty years ago, a study was made of a protein called cytochrome c, which is used by many organisms, including humans. As I shall shortly explain in detail, all proteins are made of molecular units called amino acids. Cytochrome c contains about one hundred of them, of twenty different varieties. By comparing the sequences of amino acids in cytochrome c taken from different species, we can make an estimate of the evolutionary distance they have traveled from each other. To give a concrete example, human cytochrome c is identical to that of rhesus monkeys save for a single amino acid, but there are forty-five differences between human and wheat cytochrome c. Everybody knows that humans are more closely related to monkeys than to wheat; this study shows by how much. The important point, though, is that even species as different as humans and wheat share enough structure in their respective cytochrome c molecules to confirm that we have a common ancestor, way back. Generally speaking, the farther apart two species are genetically, the longer ago they diverged on the tree of life. Unfortunately, it is not a simple matter to convert evolutionary distance into intervals of time, since mutations don't happen at a uniform rate through history. Pinpointing dates for branching events is hard.

By the late 1970s, sequencing techniques were being applied systematically to the proteins and nucleic acids of microbes as well as of higher species. Carl Woese of the University of Illinois was a pioneer in this field, and his results caused a minor sensation. Before Woese, biologists had naturally assumed that prokaryotes preceded eukaryotes by some billions of years. This would give prokaryotes pride of place at the stem of the (known) tree of life, with eukaryotes branching off, say, a billion years ago. But Woese demolished this assumption by showing that the tidy division of life into two domains—prokaryotes and eukaryotes—was fundamentally flawed. He found that there are not two, but *three* great domains of life. It turns out that the simple label "prokaryotes" encompasses two genetically quite distinct classes of cells, originally dubbed eubacteria and archaebacteria. Archaebacteria had previously been misclassified as

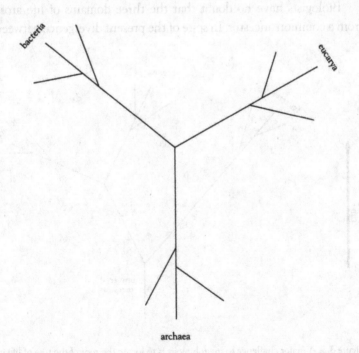

Figure 3.2. Molecular sequencing techniques show that terrestrial life is divided into three distinct domains. All multicellular organisms are restricted to the domain of eucarya.

some sort of weird strains of bacteria. Woese showed that, although archaebacteria may *look* superficially like bacteria, in terms of their biochemistry they are as different as humans and E. *coli.*

Woese's research suggested that the three basic domains—now renamed archaea, bacteria, and eucarya—split apart over three billion years ago, so the deep trifurcation of the tree of life is very old indeed, and probably occurred soon after life began (see figure 3.2). This immediately raises the important and still-unresolved question of how the three domains are situated on the tree of life. Which branched from what first? The most recent sequencing evidence, especially from the work carried out by Karl Stetter of the University of Regensburg in Germany and Norman Pace and Susan Barns of Indiana University,[3] suggests that a picture like figure 3.3 is the most likely.

Biologists have no doubt that the three domains of life arose from a common ancestor. In spite of the present divergence between

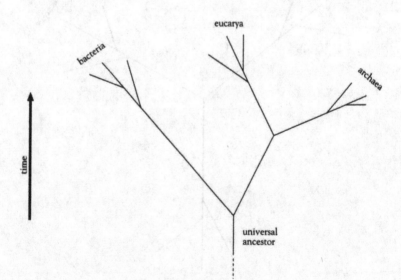

Figure 3.3. A major challenge to microbiology is to locate the root of the tree of life in relation to the trifurcation depicted schematically in figure 3.2. A likely configuration is shown above, with the universal ancestor situated somewhere between bacteria and archaea.

the domains, the basic genetic and metabolic apparatus is the same, and they share many complicated specialized features. Clearly, then, the common ancestor was already a very complex organism, not a primitive entity that had recently come into existence. I pointed out in the last section that the latest common ancestor is not the same as the first living thing. If the ancestral organism that gave rise to the three domains was already highly evolved, then it must lie well above the true base of the tree of life.

The use of molecular-sequencing techniques has revolutionized the study of microbiology and is casting new light on the mysteries of life's beginnings. In effect, it makes use of molecular fossils that lie within living cells. The results point to a very extended history for the three domains of life, with the deepest branchings occurring over three billion years ago. How well do these findings square with the more traditional method of looking for fossils in old rocks?

The earliest rock fossils

The Pilbara area of Western Australia is one of the hottest, most desolate, and least populated parts of the world. About forty kilometers to the west of the small town of Marble Bar lies the improbably named geological feature of North Pole. It was in the hilly country close to this remote spot that John Dunlop, a geology student, discovered the world's oldest known fossils in 1980. To the uninitiated, they don't look much like fossils. No ammonites or trilobites here, only some curious mounds called stromatolites. These structures are formed when cyanobacteria deposit mats of mineral grains, layer by layer, to form cushion-shaped humps. Stromatolites can still be found today in the process of formation, about five hundred kilometers from North Pole—at Shark Bay, on the Western Australian coast.[4] The fossilized stromatolites were formed in the sediment of a volcanic lagoon, and are thought to be 3.5 billion years old. Shortly after Dunlop found the Pilbara stromatolites, a group of paleontologists from California led by William Schopf discovered signs of individual fossilized microbes in rocks of a similar age in the nearby Warrawoona hills.[5] They appeared as tiny segmented filaments embedded in the chert, suggestive of cyanobacteria from some ancient sun-warmed lake.

As one heads north from the Pilbara, the nearest proper city is Darwin, named after the great scientist himself. Darwin was baffled by the apparent absence of fossils from the Precambrian era, before about six hundred million years ago. The fossils are there all right, but most Precambrian creatures were too small for their traces to be spotted by casual fossil hunters. Even experienced paleobiologists armed with state-of-the-art technology have uncovered only a few sites containing microfossils older than 2.5 billion years, and many of these finds remain contentious.

If the Warrawoona microfossils have been correctly identified as cyanobacteria, it means that life had discovered photosynthesis as early as 3.5 billion years ago. Photosynthesis is a complex and sophisticated chemical process, so it is likely that the Warrawoona micro-organisms were already fairly highly evolved, and that some more primitive precursor lived long before this epoch. Have these earlier microbes left any trace?

The chances of finding intact microfossils older than those in Western Australia are very slim. Fortunately, there are other, more subtle ways in which organisms can leave an imprint in rocks—by altering the chemical composition. For example, an early ecosystem in a shallow sea would have deposited organic material in the sediments at the bottom, creating layers of carbon-rich minerals, like a microbial graveyard. Something of this sort may have happened in the extremely old banded-iron formations at Isua in Greenland. A study of the carbon content of these rocks, pioneered by Manfred Schidlowski of the Max Planck Institute for Chemistry in Germany, hints that life may have been at work some three hundred million years before the Pilbara fossils were laid down.

The evidence for life at Isua comes from careful measurements of the isotope ratios of carbon. A typical carbon atom contains six protons and six neutrons, and for this reason it is designated C^{12}. Some carbon atoms, though, have seven neutrons instead of six, and so are designated C^{13}. Chemically, the two are identical, and are known as isotopes. Life favors C^{12}, because it is slightly lighter and reacts faster. As a result, organisms tend to sequester the lighter isotope, and thus enrich any sediments that they become entombed in. The C^{12} in the Pilbara rocks is about 3 percent above normal, and at Isua about 1 percent.

Recently a team led by Gustaf Arrhenius of the Scripps Institute of Oceanography in California used an improved technique to study the carbon-isotope ratios in the Isua rocks. Employing a device known as an ion microprobe mass spectrometer, the researchers were able to analyze tiny amounts of carbon in grains as small as ten micrometers across, weighing only twenty-trillionths of a gram, and they claim to have found an even stronger signature of life. The rocks were collected from Akilia Island, near Isua, and dated to be at least 3.85 billion years old.[6] By comparison, the age of the Earth is 4.55 billion years, as determined from radioactivity measurements. If life did exist on Earth 3.85 billion years ago, then our planet has been inhabited for at least 85 percent of its history.

Searching the fossil record might be described as a top-down approach to the investigation of biogenesis. Starting with what is known about life today, we try to follow its evolutionary path back in time, and down in size, to the simplest organisms and the earliest traces, until the record peters out in obscurity. Some time prior to 3.5 billion, and quite possibly earlier than 3.8 billion years ago, the very first terrestrial organism dwelt somewhere on our planet. But where? And what was it like? I shall address these questions when I return to the top-down route in chapter 6, but now I should like to turn to the alternative, bottom-up, approach. The idea here is to ask what we know about the conditions on the young Earth, and then try to reconstruct the physical and chemical events that sparked the beginning of life all those years ago.

Spontaneous generation

Science rejects true miracles. Although biogenesis strikes many as virtually miraculous, the starting point of any scientific investigation must be the assumption that life emerged naturally, via a sequence of normal physical processes. It is very unlikely that we will ever find out exactly what happened, but we might be able to deduce a plausible chemical pathway leading from simple chemicals to life. There may, of course, be many different pathways to life as we know it, and many alternative forms of life. It is even conceivable that scientists will one day create life of some sort in the laboratory, and thus demonstrate convincingly that a miracle isn't needed. However, in our present state of ignorance, all we can hope for are a few pointers to the key chemical steps that may have been involved. Some people may object that mere pointers are useless, and that the subject is far too speculative to be worth pursuing. That is very shortsighted. Research into the origin of life may yield valuable information even in the absence of a detailed account of how life actually began. In particular, we might be able to answer the question of just how likely or unlikely the spontaneous generation of life may be. If it turns out to be rather probable, then we can expect life to have arisen elsewhere in the universe too. On the other hand, if the chemical steps turn out to be highly improbable, then we may well be alone in the universe.

Whatever the precise chemical sequence may have been, life must have formed as a result of some sort of molecular self-assembly. The term "self-assembly" has a magical ring to it, but in fact it is very commonplace. Galaxies and crystals, for example, arise spontaneously by self-assembly; that is, they create themselves unaided from a disordered or featureless initial state. No vital forces marshal their components into the final form: only normal physical forces are involved. Biologists suppose that the same applies to the formation of life, even though the simplest form of life is immensely complicated.

As it happens, belief in the spontaneous generation of life has a long history, dating back at least to Plato. In the seventeenth century, it was widely believed that many sorts of living creatures could be generated *de novo* under appropriate conditions. Adult mice, for example, were said to appear from a heap of sweaty underwear and wheat.[7] Other favorite recipes were old socks and rotting meat from which lice, flies, and maggots might duly emerge.

Today these stories seem ridiculous, but it took a scientist of the caliber of Louis Pasteur to settle the matter. In 1862, under the incentive of a public prize, Pasteur performed a series of careful experiments to demonstrate that living organisms come only from other living organisms. A truly sterile medium would, he claimed, remain forever sterile. Pasteur declared triumphally: "Never will the doctrine of spontaneous generation recover from the mortal blow of this simple experiment!"[8]

Important though this demonstration was, Pasteur's conclusion came into direct conflict with Darwin's theory of evolution. Darwin's celebrated tome *On the Origin of Species*, which had been published just three years before Pasteur's experiments, sought to discredit the need for God to create the species by showing how one species can transmute into another. But Darwin's account left open the problem of how the *first* living thing came to exist. Unless life had always existed, at least one species—the first—cannot have come to exist by transmutation from another species, only by transmutation from nonliving matter. Darwin himself wrote, some years later: "I have met with no evidence that seems in the least trustworthy, in favour of so-called Spontaneous Generation."[9] Yet, in the absence of a miracle, life could have originated *only* by some sort of spontaneous generation. Darwin's theory of evolution and Pasteur's theory that only life begets life cannot both have been completely right.

Darwin himself was somewhat coy on the subject of life's origin (see the quotation at the start of this chapter), but he did offer the germ of an idea in a famous letter written in 1871. In this missive he envisaged "some warm little pond, with all sorts of ammonia and phosphoric salts, light, heat, electricity, etc."[10] Darwin conjectured that from this humble brew, over immense periods of time, life might form by a process of chemical complexification. This rather casual speculation was to set the trend of thinking on the subject for the next century.

At that time, the very notion that life might spring into being spontaneously from a nonliving chemical mixture was greeted with fierce criticism from theologians, and even from some scientists. The eminent British physicist Lord Kelvin dismissed the whole idea as "a very ancient speculation," opining that "science brings a vast mass of inductive evidence against this hypothesis." He stated unequivocally, "Dead matter cannot become living without coming under the influence of matter previously alive."[11] This left only two alternatives: either life has always existed or its origin was a miracle.

Little real progress was made on the subject until the 1920s, and the work of Alexander Oparin in Russia and J. B. S. Haldane in England. Both scientists recognized that it would be straining credulity to suggest that life could form suddenly, in one go, in a single amazing reaction. Taking their cue from Darwin, they assumed that a lengthy development phase must have occurred, a sequence of chemical steps gradually leading up to the first microbe. During this prebiotic stage, some yet-to-be-determined reactions successively transformed a mixture of molecules into more and more complex arrangements, until something eventually formed with the basic characteristics of a living organism.

Rather than Darwin's "little pond," Haldane envisaged the Earth's entire oceans as the setting. Rain drenching the barren landscape would have washed all manner of chemicals into the sea, there to concentrate until, to use Haldane's evocative phrase, the liquid "reached the consistency of hot dilute soup." His words were seized upon, and the description "the primordial soup" has stuck ever since.

Over the years, there have been many variations on the what and the where of the primordial soup: was it indeed an ocean, or merely a pond as Darwin thought? Could it have been a drying lagoon, a pool in a cave, or a subterranean channel? How about a boiling geyser or a volcanic vent beneath the sea? Water droplets suspended in the air? Maybe the soup wasn't located on Earth at all, but was confined to the interior of a comet or small planet. All these ideas have been seriously proposed, and most remain pure conjecture. Although the proposals differ widely, they share a common theme. They all require liquid water, laced with suitable substances, and exposed to a source of energy to drive chemical reactions forward.

Haldane and Oparin held rather different opinions about the precise sequence of events, and entrenched a schism in the subject that endures to this day. The issue concerns the formation of cells. All micro-organisms are separated from their surroundings by a membrane or cell wall. Indeed, it is hard to imagine life without a boundary of some sort. The question is, when did this cellular structure arise—before, during, or after the principal chemical steps?

Whereas Haldane focused on the chemistry of the soup, Oparin was a cells-first advocate. He was impressed by the fact that oily substances and water don't mix, and sometimes produce a suspension known as a coacervate, in which the oil retreats into tiny droplets. The oily blobs superficially resemble biological cells. Oparin's theory assumed that the physical structure of the cell came first, providing a natural containment vessel in which some molecular marvels could proceed. This idea has some attraction, because there are many physical processes (not just oil in water) that produce vesicles of some sort. Also, fluid cells and droplets can become unstable and split in two, representing a crude form of reproduction. If a bag full of chemicals swells up and undergoes fission, each of the "daughter" bags will inherit the chemical mix of the parent. This might have been enough for a rudimentary type of natural selection to operate. However, the membrane needs to have some special properties. For instance, it must trap the life-sustaining molecules inside the cell, but let through the needed raw materials from the outside.

Oparin's idea of rooting the origin of life in the formation of cells partly reflects the state of knowledge of the day. Scientists at that time were still struggling to work out the processes of metabolism and the role of proteins within the cell, with only the vaguest idea of the nature of genes. Molecular biology and knowledge of DNA did not yet exist. It was perhaps only natural that Oparin de-emphasized the genetic aspects of life and gave primacy to the physical aspects—cell formation and structure—which were better understood. That does not make the cells-first theory wrong, but it does warn us that the temptation to place the things we understand at the center of a theory risks putting the cart before the horse.

Theorizing about the origin of life seemed altogether too speculative in the 1920s, and few people paid much attention to the ideas of Oparin and Haldane. One person who did take notice, however, was Harold Urey, an American chemist who would one day win the Nobel Prize for the discovery of deuterium. Urey realized that it might be possible to test the theory of the primordial soup in the laboratory. Many years later, in 1953, he set out to do just that.

Re-creating the primordial soup

Urey's celebrated experiment was engagingly simple in conception. He sought to re-create in the lab the conditions that prevailed on the primeval Earth, and observe what happened. He had to make a guess at what the planet was like billions of years ago. The existence of liquid water was a fair bet, but the composition of the atmosphere was unknown. To decide what gases to use, Urey was guided by the fact that Earth's present atmosphere has been greatly modified by life. In particular, the oxygen content is the product of photosynthesis. So Urey excluded oxygen. This was a wise choice. Although people associate oxygen with life, it is actually a dangerously corrosive element, and a menace to most organic molecules, destroying them rapidly—as every arsonist knows. If the prebiotic phase was anything like what Haldane and Oparin had in mind, there was no free oxygen around. Urey decided on a mixture of methane, hydrogen, and ammonia.

To assist him in the experiment, Urey recruited a bright young student named Stanley Miller at the University of Chicago. Miller began by filling a glass flask with the chosen gases plus some water. He sealed the apparatus, and then passed an electric spark through the mixture to simulate the effects of sunlight. Over the next week, he watched with fascination as the water cycling through the apparatus slowly turned reddish-brown. Miller was elated: it seemed as if this simple experiment had succeeded in reproducing something like a primordial soup. Eagerly he set about analyzing the fluid and, sure enough, he found it to contain several of the organic chemicals known as amino acids, the building blocks of proteins, and basic ingredients in all terrestrial life.

Miller's intriguing results were widely hailed as the first steps on the road to the creation of life "in a test tube." If amino acids were produced in a week, it was reasoned, imagine what might happen if the experiment were continued for much longer. It might simply be a matter of time before something living crawled out of the red-brown broth. The conclusion that many scientists drew was that a few common chemicals plus an energy supply are all that is needed to create life.

Alas, the euphoria over the Miller-Urey experiment turned out to be somewhat premature, for a variety of reasons. First, geologists no longer think that the early atmosphere resembled the gas mixture in Miller's flask. The Earth probably had several different atmospheres during its first billion years, but methane and ammonia were unlikely ever to have been present in abundance. And if Earth once had substantial hydrogen in its atmosphere, it wouldn't have lasted long: being the lightest element, it would soon have escaped into space. Urey picked these gases because they all contain hydrogen. Chemists call such gases "reducing." Reduction is the opposite of oxidation, and because organics are rich in hydrogen, a reducing atmosphere is essential to produce them. However, the current best guess for the Earth's early atmosphere is that it was neither reducing nor oxidizing: rather, it was a neutral mixture of carbon dioxide and nitrogen. These gases don't readily yield amino acids.

A second reason for casting doubt on the significance of the Miller-Urey experiment is that amino acids are not all that hard to make. Many successful variants on the original Chicago setup have been tried, in which the electric spark has been replaced by a furnace, an ultraviolet lamp, shock waves, or energized chemical mixtures. It turns out that making amino acids is a cinch. In fact, they are found to occur naturally in meteorites, and even in outer space.

There is also a conceptual reason why the Miller-Urey experiment is no longer accorded its former status. It is a serious mistake to regard the road to life as a uniform highway down which a soup of chemicals is inexorably conveyed by the passage of time. Amino acids may be the building blocks of proteins, but there is a world of difference between building blocks and an assembled structure. Just as the discovery of a pile of bricks is no guarantee that a house lies around the corner, so a collection of amino acids is a long, long way from the sort of large, specialized molecules such as proteins that life requires.

Two major obstacles stand in the way of further progress towards life in a primordial soup. One is that in most scenarios the soup is far too dilute to achieve much. Haldane's vast ocean broth would be exceedingly unlikely to bring the right components together in the same place at the same time. Without some mechanism to concentrate the chemicals greatly, the synthesis of more complex substances than amino acids looks doomed. Many imaginative suggestions have been made on how to thicken the brew. For example, Darwin's warm little pond may have evaporated to leave a potent scum. Or perhaps mineral surfaces like clay trapped and concentrated passing chemicals from a fluid medium. However, it is far from clear whether any of these suggestions is realistic in the context of the early Earth, and no souplike state has been preserved in the rocks to guide us.

The other obstacle is even deeper and goes back to the second law of thermodynamics. Recall how this law describes a natural tendency towards degradation and corruption, and away from increasing order and complexity. The synthesis of complex biomolecules therefore runs "against the tide," thermodynamically speaking. At first sight this seems to lead to a contradiction, because amino acids form readily under a wide range of conditions. In fact, there is no conflict with the second law. As I explained in chapter 2, order can appear in one place as long as a greater quantity of disorder, or entropy, is delivered to the environment. This is what happens when a crystal forms from a solute. The crystalline solid is a more ordered arrangement of atoms than a liquid, so it has less entropy. However, the formation of a crystal is accompanied by a release of heat into the environment, which raises the entropy. The second factor out-weighs the first. The same applies to amino-acid synthesis, which, like crystal formation, is thermodynamically favored. The reason for this concerns the role of free energy. If a process lowers the energy of a system,—i.e., if it goes "downhill"—then it has the second law's blessing. By contrast, an "uphill" process defies the second law. Wa-ter runs downhill, not uphill. You *can* make water go uphill, but only if you work for it. A process that happens spontaneously is always a downhill process. Amino-acid production has this character of be-ing a downhill process, which is why amino acids are so easy to make.

But now we hit a snag. The second step on the road to life, or at least the road to proteins, is for amino acids to link together to form molecules known as peptides. A protein is a long peptide chain, or a polypeptide. Whereas the spontaneous formation of amino acids from an inorganic chemical mixture is an allowed downhill process, coupling amino acids together to form peptides is an uphill process. It therefore heads in the wrong direction, thermodynamically speaking. Each peptide bond that is forged requires a water molecule to be plucked from the chain. In a watery medium like a primordial soup, this is thermodynamically unfavorable. Consequently, it will not happen spontaneously: work has to be done to force the newly extracted water molecule into the water-saturated medium. Obvi-ously peptide formation is not impossible, because it happens inside living organisms. But there the uphill reaction is driven along by the use of customized molecules that are pre-energized to supply the necessary work. In a simple chemical soup, no such specialized mol-ecules would be on hand to give the reactions the boost they need. So a watery soup is a recipe for molecular disassembly, not self-assembly.[12]

To be sure, there would have been no lack of available energy sources on the early Earth to provide the work needed to forge the peptide bonds, but just throwing energy at the problem is no solu-tion. The same energy sources that generate organic molecules also serve to destroy them. To work constructively, the energy has to be targeted at the specific reaction required. Uncontrolled energy input, such as simple heating, is far more likely to prove destructive than constructive. The situation can be compared to a workman laboriously building a brick pillar by piling bricks one on top of another. The higher the pillar goes, the more likely it is to wobble and collapse. Likewise, long chains made of amino acids linked together are very fragile. As a general rule, if you simply heat organics willy-nilly, you end up, not with delicate long chain molecules, but with a tarry mess, as barbecue owners can testify.

It is true that the second law of thermodynamics is only a statistical law; it does not absolutely forbid physical systems from going "the wrong way" (i.e., uphill). But the odds are heavily weighted against it. So, for example, it is possible, but very unlikely, to create a brick pillar by simply tipping a pile of bricks out from a hopper. You might not be surprised to see two bricks ending up neatly on top of one another; three bricks would be remarkable, ten almost miraculous. You would undoubtedly wait a very long time for a ten-brick column to happen spontaneously. In ordinary chemical reactions that take place close to thermodynamic equilibrium, the molecules are jiggled about at random, so again you will likely wait a very long time for a fragile molecular chain to form by accident. The longer the chain, the longer the wait. It has been estimated that, left to its own devices, a concentrated solution of amino acids would need a volume of fluid the size of the observable universe to go against the thermodynamic tide and create a single small polypeptide spontaneously. Clearly, random molecular shuffling is of little use when the arrow of directionality points the wrong way.

One possible escape route from the strictures of the second law is to depart from thermodynamic-equilibrium conditions. The American biochemist Sidney Fox has investigated what happens when a mixture of amino acids is strongly heated. Driving out the water as steam makes the linkage of amino acids into peptide chains much more likely. The thermal-energy flow generates the necessary entropy to comply with the second law. Fox has produced some quite long polypeptides, which he terms "proteinoids," using this method. Unfortunately, the resemblance between Fox's proteinoids and real proteins is rather superficial. For example, real proteins are made exclusively of left-handed amino acids (see page 71), whereas proteinoids are an equal mixture of left and right.

There is a more fundamental reason why the random self-assembly of proteins seems a nonstarter. This has to do not with the formation of the chemical bonds as such, but with the particular order in which the amino acids link together. Proteins do not consist of any old peptide chains; they are very specific amino-acid sequences that have specialized chemical properties needed for life. However, the number of alternative permutations available to a mixture of amino acids is superastronomical. A small protein may typically contain a hundred amino acids of twenty varieties. There are about 10^{130} (which is one followed by 130 zeros) different arrangements of the amino acids in a molecule of this length.[13] Hitting the right one by accident would be no mean feat.[14]

Getting a useful configuration of amino acids from the squillions of useless combinations on offer can be thought of as a mammoth information-retrieval problem, like trying to track down a site on the Internet without a search engine. The difficulty can be expressed in thermodynamic terms by recalling the connection between information and entropy explained in chapter 2. The highly special information content of a protein represented by its very specific amino-acid sequence implies a big decrease in entropy when the molecule forms. Again, the mere uncontrolled injection of energy won't accomplish the ordered result needed. To return to the bricklaying analogy, making a protein simply by injecting energy is rather like exploding a stick of dynamite under a pile of bricks and expecting it to form a house. You may liberate enough energy to raise the bricks, but, without coupling the energy to the bricks in a controlled and ordered way, there is little hope of producing anything other than a chaotic mess. So making proteins by randomly shaking amino acids runs into double trouble, thermodynamically. Not only must the molecules be shaken "uphill," they have to be shaken into a configuration that is an infinitesimal fraction of the total number of possible combinations.

So far I have just been talking about making proteins by linking amino acids into peptides. But proteins are only a small part of the intricate fabric of life. There are lipids and nucleic acids and ribosomes, and so on. And here we hit yet another snag. It is possible that scientists, using complicated and delicate laboratory procedures, may be able to synthesize piecemeal the basic ingredients of life. What is far less likely is that the same set of procedures would yield all the required pieces at the same time. Thus, not only is there a mystery about the self-assembly of large, delicate, and very specifically structured molecules from an incoherent mêlée of bits, there is also the problem of producing, simultaneously, a collection of many different types of molecules.

Let me spell out what is involved here. I have already emphasized that the complex molecules found in living organisms are not themselves alive. A molecule is a molecule is a molecule; it is neither living nor dead. Life is a phenomenon associated with a whole society of specialized molecules, millions of them, cooperating in surprising and novel ways. No single molecule carries the spark of life, no chain of atoms alone constitutes an organism. Even DNA, the biological supermolecule, is not alive. Pluck the DNA from a living cell and it would be stranded, unable to carry out its familiar role. Only within the context of a highly specific molecular milieu will a given molecule play its role in life. To function properly, DNA must be part of a large team, with each molecule executing its assigned task alongside the others in a cooperative manner.

Acknowledging the interdependability of the component molecules within a living organism immediately presents us with a stark philosophical puzzle. If everything needs everything else, how did the community of molecules ever arise in the first place? Since most large molecules needed for life are produced only by living organisms, and are not found outside the cell, how did they come to exist originally, without the help of a meddling scientist? Could we seriously expect a Miller-Urey type of soup to make them all at once, given the hit-and-miss nature of its chemistry?

You might get the impression from what I have written not only that the origin of life is virtually impossible, but that life itself is impossible. If fragile biomolecules are continually being attacked and disintegrated, surely our own bodies would rapidly degenerate into chemical mayhem spelling certain death? Fortunately for us, our cells contain sophisticated chemical-repair-and-construction mechanisms, handy sources of chemical energy to drive processes uphill, and enzymes with special properties that can smoothly assemble complex molecules from fragments. Also, proteins fold into protective balls that prevent water from attacking their delicate chemical bonds. As fast as the second law tries to drag us downhill, this cooperating army of specialized molecules tugs the other way. As long as we remain open systems, exchanging energy and entropy with our environment, the degenerative consequences of the second law can be avoided. But the primordial soup lacked these convenient cohorts of cooperating chemicals. No molecular-repair gangs stood ready to take on the second law. The soup had to win the battle alone, against odds that were not just heavy but mind-numbingly huge.

So what is the answer? Is life a miracle after all? In chapter 4, I shall look carefully at the latest attempts to explain how a chemical mixture might effectively shorten the vast odds stacked up against the spontaneous assembly of complex molecules. Here I just want to make a general point. The first living things were undoubtedly far more primitive than today's microbes. You can't look at extant bacteria, with their fine-tuned and specialized metabolisms, and expect all their components to have been made and assembled in their existing form in a primordial soup. Today's microbes have emerged only gradually, after a long period of evolutionary refinement, from rough-and-ready beginnings. Early life would have been far sloppier, biochemically, than today's organisms.

This illustrates an important general principle: crude machines are more robust than sophisticated ones. The more finessed a machine is, the more vulnerable its components become. Pour crude diesel fuel into the tank of a finely tuned racing car and it will cough and splutter ineffectually, but a tractor can chug away contentedly.

By the same token, if you were to drop a modern DNA molecule in the primordial soup, it would be rendered helpless. But a less refined precursor of DNA might fare better and replicate successfully. It seems that life must have begun as a ramshackle process and became refined and streamlined over time. Perhaps the odds against the self-assembly of the microbial equivalent of a tractor are not insurmountable.

Chance and the origin of life

Ask the simple question: Given the conditions that prevailed on the Earth four billion years ago, how likely was it that life arose?

The following answer won't do: "Life was inevitable, because we are here." Obviously life *did* originate—our existence proves that much. But did it *have* to originate? In other words, was the emergence of life from a chemical broth or whatever inevitable, given millions of years?

Nobody knows the answer to this question. The origin of life may have been a sheer fluke, a chemical accident of stupendous improbability, an event so unlikely that it would never happen twice in the entire universe. Or it may have been as unremarkable and predetermined as the formation of salt crystals. How can we know which explanation is the right one?

Let's take a look at the chemical-fluke theory. As explained earlier in this chapter, terrestrial life is based on some very complicated molecules with carefully crafted structures. Even in simple organisms, DNA contains millions of atoms. The precise sequence of atoms is crucial. You can't have an arbitrary sequence, because DNA is an instruction manual for making the organism. Change a few atoms and you threaten the structure of the organism. Change too many and you won't have an organism at all.

The situation may be compared to the word sequence of a novel. Change a few words here and there at random, and the text will probably be marred. Scramble all the words and there is a very high probability that it won't be a novel any more. There will be other novels with similar words in different combinations, but the set of word sequences that make up novels is an infinitesimal fraction of all possible word sequences.

In the previous section I presented the fantastic odds against shuffling amino acids at random into the right sequence to form a protein molecule by accident. That was a single protein. Life as we know it requires hundreds of thousands of specialist proteins, not to mention the nucleic acids. The odds against producing just the proteins by pure chance are something like $10^{40\,000}$ to 1. This is one followed by forty thousand zeros, which would take up an entire chapter of this book if I wanted to write it out in full. Dealing a perfect suit at cards a thousand times in a row is easy by comparison. In a famous remark, the British astronomer Fred Hoyle likened the odds against the spontaneous assembly of life to those for a whirlwind sweeping through a junkyard and producing a fully functioning Boeing 747.[15]

I often give public lectures on the possibility of extraterrestrial life. Invariably, someone in the audience will remark that there *must* be life on other planets because there are so many stars offering potential abodes. It is a commonly used argument. On a recent trip to Europe to attend a conference on extraterrestrial life, I flipped through the airline's in-flight entertainment guide, only to find that the search for life beyond Earth was on offer as part of their program. The promotional description said "With a half-trillion stars wheeling through the spiral patterns of the Milky Way Galaxy, it seems illogical to assume that among them only one world supports intelligent life."[16] The use of the word "illogical" was unfortunate, because the logic is perfectly clear. There are indeed a lot of stars— at least ten billion billion in the observable universe. But this number, gigantic as it may appear to us, is nevertheless *trivially* small compared with the gigantic odds against the random assembly of even a single protein molecule. Though the universe is big, if life formed solely by random agitation in a molecular junkyard, there is scant chance it has happened twice.

Some people feel that something as basic as our own existence can't be put down to a chemical quirk, and that sweeping the problem under the carpet with the word "accident" is a cop-out. Sometimes the principle of mediocrity is cited: There is nothing special or exceptional about our place in the universe. The Earth appears to be a typical planet around a typical star in a typical galaxy. So why should life on Earth not also be typical?

Unfortunately, this argument won't wash. Our own existence must be the exception to the rule that what we observe is unexceptional. If there is only one planet in the universe with life, it has to be ours! Obviously we won't find ourselves living on a lifeless planet, by definition. So Earth will not be a randomly selected planet in a cosmic sample, because we have selected it by our very existence.

In spite of this undeniable fact, scientists should attempt to explain the world in terms of laws and principles wherever possible. You'd never get away with arguing that the rings of Saturn formed as an accidental association of independently moving particles. Resorting to flukes must be seen as a last resort. That doesn't mean flukes never happen, or might not be important.[17] It may be that life on Earth is a fluke. But we should at least make an attempt to explain biogenesis as a normal physical process. In the coming chapters, I shall look at some suggestions for shortening the apparently colossal odds against the spontaneous genesis of life.

CHAPTER 4

The Message in the Machine

IN JULY 1997, scientists at Cornell University released photographs of a guitar no larger than a human blood cell. Its strings are just one hundred atoms thick. This Lilliputian instrument was sculpted from crystalline silicon, using an etching technique involving a beam of electrons. It was intended as a gimmick, but it dramatically illustrated an important technological development: machines can now be made that are too small to be seen with the naked eye. Scientists have fabricated invisible cogwheels, motors the size of a pinhead, and electrical switches as tiny as individual molecules. Engineers at IBM have even been able to imprint the company's initials atom by atom on a crystal surface. The burgeoning field of nanotechnology—building structures and devices measured on a scale of billionths of a meter—promises to revolutionize our lives.

These achievements of microengineering are breathtaking in their implications, but we should not lose sight of the fact that nature got there first. The world is already full of nanomachines: they are called living cells. Each cell is packed with tiny structures that might have come straight from an engineer's manual. Minuscule tweezers, scissors, pumps, motors, levers, valves, pipes, chains, and even vehicles abound. But of course the cell is more than just a bag of gadgets. The various components fit together to form a smoothly functioning whole, like an elaborate factory production line. The miracle of life is not that it is made of nanotools, but that these tiny diverse parts are integrated in a highly organized way.

What is the secret of this astonishing organization? How can stupid atoms do it? Individually, atoms can only jostle their neighbors and bond to them if the circumstances are right. Yet, collectively, they accomplish ingenious marvels of construction and control, with a fine-tuning and complexity as yet unmatched by any human engineering. Somehow nature discovered, on its own, how to do this. It found out how to build the intricate machine we call the living cell, using only the raw materials to hand, all jumbled up. It repeats this feat every day in our own bodies, every time a new cell is made. That is already a fantastic accomplishment. Even more remarkable is that nature built the first cell from scratch. How was it done?

As a simple-minded physicist, when I think about life at the molecular level, the question I keep asking is: how do all these mindless atoms know what to do? The complexity of the living cell is immense, resembling a city in the degree of its elaborate activity. Each molecule has a specified function and a designated place in the overall scheme so that the correct objects are manufactured. There is much commuting going on. Molecules have to travel across the cell to meet others at the right place and the right time in order to carry out their jobs properly. This all happens without a boss to order the molecules around and steer them to their appropriate locations. No overseer supervises their activities. Molecules simply do what molecules have to do: bang around blindly, knock into each other, rebound, embrace. At the level of individual atoms, life is anarchy—blundering, purposeless chaos. Yet somehow, collectively, these unthinking atoms get it together and perform the dance of life with exquisite precision.

Can science ever explain such a magnificently self-orchestrating process? Some people flatly deny it.[1] They believe that the living cell is just too elaborate, too contrived, to be the product of blind physical forces alone. Science may give a good account of this or that individual feature, they say, but it will never explain the overall organization, or how the original cell was assembled in the first place.

I beg to differ. Science will, I believe, eventually give a convincing explanation for the origin of life, but only if the problem is tackled on two levels. The first is the molecular level, the subject of this chapter. This is where progress has been most impressive. Over the past few decades, molecular biology has made gigantic strides elucidating which molecules do what to which. Always it is found that nature's nanomachines operate according to perfectly ordinary physical forces and laws. No weird goings-on have been discovered. It would be wrong, however, to suppose that molecules are all that there is to life. We no more explain life by cataloguing its molecular activities than we account for the genius of Mozart or Einstein by determining how a neuron works. To use the cliché, the whole is more than the sum of its parts. The very word "organism" implies cooperation at a global level that cannot be captured in the study of the components alone. Without understanding its collective activity, the job of explaining life is only partly done.

Replicate, replicate!

In chapter 1, I put reproduction near the top of my list of defining properties of life. Without it, and in the absence of immortality, all life would sooner or later cease. For a long time scientists had very little idea how organisms reproduce themselves. Vague notions of invisible genes conveying biological messages from one generation to the next revealed little of how cells actually do it. With the advent of molecular biology and the discovery of DNA, however, the mystery was finally solved.

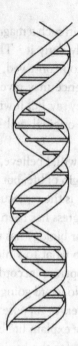

Figure 4.1. A double helix. The structure of the DNA molecule is shown here schematically. Note that the twin helical strands are connected by cross-links. These play the crucial role in storing biological information.

Boiled down to its essentials, the secret of reproduction lies with molecular replication. The idea of a molecule making a copy of itself may seem rather magical, but it actually turns out to be quite straightforward. The principle underlying it is, in fact, an exercise in elementary geometry. The first point to grasp may be obvious, but it is crucially important: molecules have definite shapes. Organic molecules are not simple ball-like blobs; they boast all sorts of appendages, such as arms, elbows, cavities, and rings. Although the interatomic forces dictate what sticks to (or repels) what, it is the overall three-dimensional structure of organic molecules that largely determines, Lego-like, their biological capabilities. The Pythagorean philosophers, who believed that geometry was the key to the universe, would have been delighted.

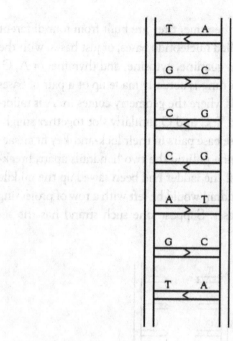

Figure 4.2. Unraveled DNA. Here the double helix is smoothed out to reveal the form of a ladder. The rungs of the ladder are made up of complementary pairs of molecules shaped to fit snugly like a lock and key.

DNA is the genetic databank, and the replication of this macromolecule lies at the heart of biological reproduction. Let me describe how DNA sets about copying itself, using simple geometry. The structure of DNA is the famous double helix, discovered by Crick and Watson in the early 1950s. Its form is shown schematically in figure 4.1. Notice that the two helical strands are attached by cross-links. The helical shape is incidental to my explanation, so, to make things simpler, imagine unwinding the entwined coils and laying them out to make a ladder (see figure 4.2). The handrails of the ladder are the two unwound helices, and the rungs correspond to the cross-links. The handrails perform a purely scaffolding role, holding the molecule together. The business part of DNA lies with the rungs.

The rungs aren't all the same: they are built from four different varieties of molecules called nucleotide bases, or just bases, with the chemical names adenine, guanine, cytosine, and thymine, or A, G, C, and T for short. Each rung is actually made up of a pair of bases joined end to end. This is where the geometry comes in. A is tailor-made to butt neatly with T; C and G similarly slot together snugly. The forces that bind these base pairs in their lock-and-key fit are actually rather weak. Imagine pulling the two handrails apart, breaking all the base pairs, as if the ladder had been sawed up the middle (see figure 4.3). Each handrail would be left with a row of projecting arms—the unmarried bases. Suppose one such strand has the se-

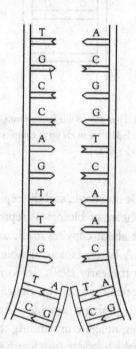

Figure 4.3. Replication: the key property of life. If the cross-links of the DNA molecule come apart, the projecting stumps can attract the appropriate individual bases that might be milling about to build a new complementary strand. When each half of the original DNA molecule has done this, the molecule will have replicated itself.

quence TGCCAGTT . . . ; then the opposing strand would have the complementary sequence ACGGTCAA. . . . You could reassemble the ladder by lining up the appropriate base pairs again and snapping all the complementary open ends together. The fact that every base along the DNA molecule is partnered in this way makes each strand a template for the other. If you have only one strand, no worries; you can figure out the structure of the other by using the base-pairing rules: A with T, C with G.

It is this templating, or complementarity, that is the basis for the replication process. To see how, imagine that some of the double helix is unzipped, as described above, leaving a run of unattached bases sticking out of each strand. If there is a supply of free base molecules—A's, G's, C's, and T's—floating around, then they will tend to slot in and stick to these exposed stumps—A to T, C to G, T to A, and G to C—and thereby automatically reconstruct a new complementary strand. As long as the base-pairing rules work correctly, the new strand is guaranteed to be identical to the original. So, if a DNA molecule is pulled apart, each exposed strand will build onto itself a new partner strand, thus making two DNA molecules in place of one. Note that this sort of template replication, where the strands are complementary, is not really like Xeroxing, but more analogous to photographic reproduction using a negative.

Thus the structural replication of DNA is readily explained. But this still leaves the question of genes and heredity. How does DNA store and transmit genetic information? This is where the four different bases come in. You can think of A, G, C, and T as a four-letter alphabet. The precise sequence of letters can be used to spell out a message. A gene is simply a long string of base pairs, or letters, conveying part of this message. When the DNA replicates, an identical sequence is built into the duplicate. Because of the double-stranded, complementary nature of this process, each DNA molecule actually contains two copies of the message, a positive and a negative, so all the information needed to make a complete DNA molecule is contained in either of its strands.

The replication process works very effectively, with the help of some specialized enzymes that facilitate the unzipping and joining operations. Just how effectively is evident from the fact that DNA in its basic structure has survived for over three billion years. However, no copying process is perfect, and it is inevitable that errors will creep in from time to time. These will alter the sequence of bases—i.e., scramble up the letters A, G, C, and T. Because DNA is a recipe for making an organism, if the message gets a bit garbled during replication, the resulting organism may suffer a mutation. The copying errors are the source of variation between generations that natural selection exploits. The genetic messages are impressively long. A simple bacterium like *E. coli* contains a few million symbols in its genome (a genome is a complete set of genes), enough to fill a thousand-page book. Human DNA would require a whole library. As I explained in chapter 1, DNA contains the total information needed to build and operate the organism to which it belongs. Viewed like this, life is just a string of four-letter words.

Making a living

So far, I have made life seem to be all about DNA, genes, and replication. It's true that, in a narrow biological sense, life is simply in the business of replicating genes. But DNA is helpless on its own. It must build a cell, with all its specialized chemicals, actually to effect the replication process. In so-called higher life forms, it must build a whole organism in order to replicate. From the perspective of a genome, an organism is a roundabout way of copying DNA.

Why do genes need organisms to lend a hand? Why can't they just replicate on their own? The answer is: because organisms can *do* things, such as move away from danger and forage for raw materials. This helps the DNA to replicate more efficiently. But building biomass and doing things needs other stuff; DNA is absolutely no good for those purposes. That other stuff comes mainly in the form of proteins, the second important class of specialized organic molecules. As I have already remarked, life as we know it is the upshot of a mutually beneficial deal struck between DNA and proteins.

Proteins are a godsend to DNA, because they can be used both as building material, to make things like cell walls, and as enzymes, to supervise and accelerate chemical reactions. Enzymes are chemical catalysts that "grease the wheels" of the biological machine. Without them metabolism would grind to a halt, and there would be no energy available for the business of life. Not surprisingly, therefore, a large part of the DNA databank is used for storing instructions on how to make proteins.

Here is how those instructions get implemented. Remember that proteins are long chain molecules made from lots of amino acids strung together to form polypeptides. Each different sequence of amino acids yields a different protein. The DNA has a wish list of all the proteins the organism needs. This information is stored by recording the particular amino-acid sequence that specifies each and every protein on the list. It does so using DNA's four-letter alphabet, A, G, C, and T; the exact sequence of letters spells out the amino-acid recipe, protein by protein—typically a few hundred base pairs for each.

To turn this dry list of amino acids into assembled, functioning proteins, DNA enlists the help of a closely related molecule known as RNA (for ribonucleic acid). RNA is also made from four bases, A, G, C, and U. Here U stands for uracil; it is similar to T and serves the same purpose alphabetically. RNA comes in several varieties; the one of interest to us here is known as messenger RNA, or mRNA for short. Its job is to read off the protein recipes from DNA and convey them to tiny factories in the cell where the proteins are made. These mini-factories, called ribosomes, are complicated machines built from RNA and proteins of various sorts. Ribosomes come with a slot into which the mRNA feeds, after the fashion of a punched tape of the sort used by old-fashioned computers. The mRNA "tape" chugs through the ribosome, which then carries out its instructions bit by bit, hooking amino acids together, one by one in the specified sequence, until an entire protein has been constructed. Earthlife makes proteins from twenty different varieties of amino acids,[2] and the mRNA records which one comes after which so the ribosome can put them together in the right order.

It is quite fascinating to see how the ribosome goes about joining the amino acids up into a chain. Naturally the amino acids don't obligingly come along in the right order, ready to be hooked onto the end of the chain. So how does the ribosome ensure that the mRNA gets its specified amino acid at each step? The answer lies with another set of RNA molecules, called transfer RNA, or tRNA for short. Each particular tRNA molecule brings along to the ribosome factory one and only one sort of amino acid stuck to its end, to present it to the production line.

At each step in the assembly of the protein, the trick is to get the *right* tRNA, with the right amino acid attached, to give up its cargo and transfer it to the end of the growing protein chain, while rejecting any of the remaining nineteen alternatives that may be on offer. This is accomplished as follows. The mRNA (remember, this carries the instructions) exposes a bit of information (i.e., a set of "letters") that says, "Add amino acid such-and-such now." The instructions are implemented correctly because only the targeted tRNA molecule, carrying the designated amino acid, will recognize the exposed bit of mRNA from its shape and chemical properties, and bind to it. The other tRNA molecules—the ones that are carrying the "wrong" amino acids—won't fit properly into the binding site. Once the right tRNA molecule has been seduced into berthing at the production line, the next step is for the ribosome to persuade the newly arrived amino-acid cargo to attach itself to the end of the protein chain. The chain is waiting in the ribosome, dangling from the end of the previously selected tRNA molecule. At this point the latter molecule lets go and quits the ribosome, passing the entire chain on to the newly arrived tRNA, where it links onto the amino acid it has brought with it. The chain thus grows by adding amino acids to the head rather than the tail. If you didn't follow all of this on the first read-through, don't worry: it isn't essential for understanding what follows. I just thought it was sufficiently amazing to be worth relating in some detail.

When the protein synthesis is complete, the ribosome receives a "stop" signal from the mRNA "tape" and the chain cuts loose. The protein is now assembled, but it doesn't remain strung out like a snake. Instead it rolls up into a knobbly ball, rather like a piece of elastic that's stretched and allowed to snap back. This folding process may take some seconds, and it is still something of a mystery how the protein attains the appropriate final shape. If it is to work properly, the three-dimensional form of the protein has to be correct, with the bumps and cavities in all the right places, and the right atoms facing outwards. Ultimately, it is the particular amino-acid sequence along the chain that determines the final three-dimensional conformation, and therefore the physical and chemical properties, of the protein.

This whole remarkable sequence of events is repeated in thousands of ribosomes scattered throughout the cell, producing tens of thousands of different proteins. It is worth repeating that, in spite of the appearance of purpose, the participating molecules are completely mindless. Collectively they may display systematic cooperation, as if to a plan, but individually they just career about. The molecular traffic within the cell is essentially chaotic, driven by chemical attraction and repulsion and continually agitated by thermal energy. Yet out of this blind chaos order emerges spontaneously.

The above account, thrilling though it may be, may give the impression that, replication apart, fabricating proteins is all there is to life. Indeed, it is easy to gain this impression from reading molecular-biology textbooks. However, "Make proteins!" is a pretty thin job description for DNA. Surely there is more to life than that? What about mating rituals, nest building, and social structure? Dazzling behavioral feats like bird migration or the weaving of spider webs?

To comprehend life in all its magnificent complexity means going beyond mere molecules and considering the organism as a whole, with its hierarchy of levels and large-scale organization. It also requires distinguishing between structure and function. The success of molecular biology stems in large part from the elucidation of the shapes and chemical affinities of certain molecules such as bases and proteins. But life cannot be reduced to a collection of static shapes thrown haphazardly together. The organizational power of living things requires cooperative *processes* that encompass many molecules and integrate their behavior into a coherent unity. So something crucial has been left out of the account so far. What is it?

The answer is buried in the foregoing description of protein production. I began by explaining the geometrical forms of molecules, the structure of DNA, and the sequence of base pairs; then I sneakily started describing messages and information and specifications. In short, I shifted from the language of hardware to that of software. A gene is a particular material form in three-dimensional space, but it is also an instruction to do something. The secret of life lies with this dual function of biological components. And nothing better illustrates this duality than the genetic code.

The genetic code

I have described life as a deal struck between nucleic acids and proteins. However, these molecules inhabit very different chemical realms; indeed, they are barely on speaking terms. This is most clearly reflected in the arithmetic of information transfer. The data needed to assemble proteins are stored in DNA using the four-letter alphabet A, G, C, T. On the other hand, proteins are made out of twenty different sorts of amino acids. Obviously twenty into four won't go. So how do nucleic acids and proteins communicate?

Earthlife has discovered a neat solution to this numerical mismatch by packaging the bases in triplets. Four bases can be arranged in sixty-four different permutations of three, and twenty *will* go into sixty-four, with some room left over for redundancy and punctuation. The sequence of rungs of the DNA ladder thus determines, three by three, the exact sequence of amino acids in the proteins.

To translate from the sixty-four triplets into the twenty amino acids means assigning each triplet (termed a codon) a corresponding amino acid. This assignment is called the genetic code. The idea that life uses a cipher was first suggested in the early 1950s by George Gamow, the same physicist who proposed the modern big-bang theory of the universe. As in all translations, there must be someone, or something, that is bilingual, in this case to turn the coded instructions written in nucleic-acid language into a result written in amino-acid language. From what I have explained, it should be apparent that this crucial translation step occurs in living organisms when the appropriate amino acids are attached to the respective molecules of tRNA prior to the protein-assembly process. (Sorry, you might have to go back and read pages 105–106 after all.) This attachment is carried out by a group of clever enzymes that recognize *both* RNA sequences *and* the different amino acids, and marry them up accordingly with the right designation.

The genetic code, with a few recently discovered minor variations, is common to all known forms of life. That the code is universal is extremely significant, for it suggests it was used by the common ancestor of all life, and is robust enough to have survived through billions of years of evolution. Without it, the production of proteins would be a hopelessly hit-or-miss affair.

Questions abound. How did such a complicated and specific system as the genetic code arise in the first place? Why, out of the 10^{70} possible codes based on triplets, has nature chosen the one in universal use? Could a different code work as well? If there is life on Mars, will it have the same genetic code as Earthlife? Can we imagine uncoded life, in which interdependent molecules deal directly with each other on the basis of their chemical affinities alone? Or is the origin of the genetic code itself (or at least *a* genetic code) the key to the origin of life? The British biologist John Maynard Smith has described the origin of the code as the most perplexing problem in evolutionary biology. With collaborator Eörs Szathmáry he writes: "The existing translational machinery is at the same time so complex, so universal, and so essential that it is hard to see how it could have come into existence, or how life could have existed without it."[3]

To get some idea of why the code is such an enigma, consider whether there is anything special about the numbers involved. Why does life use twenty amino acids and four nucleotide bases? It would be far simpler to employ, say, sixteen amino acids and package the four bases into doublets rather than triplets. Easier still would be to have just two bases and use a binary code, like a computer. If a simpler system had evolved, it is hard to see how the more complicated triplet code would ever take over. The answer could be a case of "It was a good idea at the time." If the code evolved at a very early stage in the history of life, perhaps even during its prebiotic phase, the numbers four and twenty may have been the best way to go for chemical reasons relevant at that stage. Life simply got stuck with these numbers thereafter, their original purpose lost. Or perhaps the use of four and twenty is the optimum way to do it. There is an advantage in life's employing many varieties of amino acid, because they can be strung together in more ways to offer a wider selection of proteins. But there is also a price: with increasing numbers of amino acids, the risk of translation errors grows. With too many amino acids around, there would be a greater likelihood that the wrong one would be hooked onto the protein chain. So maybe twenty is a good compromise.

An even tougher problem concerns the coding assignments— i.e., which triplets code for which amino acids. How did these designations come about? Because nucleic-acid bases and amino acids don't recognize each other directly, but have to deal via chemical intermediaries, there is no obvious reason why particular triplets should go with particular amino acids. Other translations are conceivable. Coded instructions are a good idea, but the actual code seems to be pretty arbitrary. Perhaps it is simply a frozen accident, a random choice that just locked itself in, with no deeper significance. On the other hand, there may be some subtle reason why this particular code works best. If one code had the edge over another, reliability-wise, then evolution would favor it, and, by a process of successive refinement, an optimal code would be reached. It seems reasonable. But this theory is not without problems either. Darwinian evolution works in incremental steps, accumulating small advantages over many generations. In the case of the code, this won't do. Changing even a single assignment would normally prove lethal, because it alters not merely one but a whole set of proteins. Among these are the proteins that activate and facilitate the translation process itself. So a change in the code risks feeding back into the very translation machinery that implements it, leading to a catastrophic feedback of errors that would wreck the whole process. To have accurate translation, the cell must first translate accurately.

This conclusion seems paradoxical. A possible resolution has been suggested by Carl Woese.[4] He thinks the code assignments and the translation mechanism evolved together. Initially there was only a rough-and-ready code, and the translation process was very sloppy. At this early stage, which is likely to have involved less than the present complement of twenty amino acids, organisms had to make do with very inefficient enzymes: the highly specific and refined enzymes life uses today had not yet evolved. Obviously some coding assignments would prove better than others, and any organism that employed the least error-prone assignments to code for its most important enzymes would be on to a winner. It would replicate more accurately, and in the process its coding arrangements would predominate among daughter cells. In this context, a "better" coding assignment would mean a robust one, so that, if there was a translation error, the same amino acid would nevertheless be made—i.e., there would be enough ambiguity for the error to make no difference. Or, in case the error did cause a different amino acid to be made, it would be a close cousin of the intended one, and the resulting protein would do the job almost as well. Successive refinements of this process might then lead to the universal code seen today—like a picture gradually coming into focus.

The code may have an altogether deeper explanation. If a table of coding assignments is drawn up, it can be analyzed mathematically to see if there are any hidden patterns. Peter Jarvis and his colleagues at the University of Tasmania claim that the universal code conceals abstract sequences similar to the energy levels of atomic nuclei, and might even involve a subtle property of subatomic particles called supersymmetry.[5] These mathematical correspondences may be purely coincidental, or they may point to some underlying connection between the physics of the molecules involved and the organization of the code.

I have subjected the reader to the technicalities of the genetic code to make a general conceptual point that goes right to the heart of the mystery of life. Any coded input is merely a jumble of useless data unless an interpreter or a key is available. A coded message is only as good as the context in which it is put to use. That is to say, it has to *mean* something. In chapter 2, I drew the distinction between syntactic and semantic information. On their own, genetic data are mere syntax. The striking utility of encoded genetic information stems from the fact that amino acids "understand" it. The information distributed along a strand of DNA is *biologically relevant*. In computerspeak, genetic data are *semantic* data.

For a clear perspective on this point, consider the way in which the four bases A, G, C, and T are arranged in DNA. As explained, these sequences are like letters in an alphabet, and the letters may spell out, in code, the instructions for making proteins. A different sequence of letters would almost certainly be biologically useless. Only a very tiny fraction of all possible sequences spells out a biologically meaningful message, in the same way that only certain very special sequences of letters and words constitute a meaningful book.[6] Another way of expressing this is to say that genes and proteins require exceedingly high degrees of specificity in their structure. As I stated in my list of properties in chapter 1, living organisms are mysterious not for their complexity *per se*, but for their tightly specified complexity. To comprehend fully how life arose from nonlife, we need to know not only how biological information was concentrated, but also how biologically useful information came to be *specified*, given that the milieu from which the first organism emerged was presumably just a random mix of molecular building blocks. In short, how did meaningful information emerge spontaneously from incoherent junk?

I began this section by stressing the dual nature of biomolecules: they can be both hardware—particular three-dimensional forms—and software. The genetic code shows just how important the informational aspect of biomolecules is. The job of explaining the origin of life goes beyond finding a plausible chemical pathway out of a primordial soup. We need to know, conceptually, how mere hardware can give rise to software.

Getting the message

I am writing this book on an old-fashioned Macintosh Classic computer, with a small screen and an intriguing habit of choosing its own tab setting. Like most computers, my Mac is made mainly from plastic, but the crucial innards consist of metal and semiconductors. Together with the wires, circuit boards, and glass screen, this constitutes the computer hardware. The machine is useless, however, without the software that instructs it on what to do. Mostly the software comes loaded on floppy disks. Of course the disks are also hardware, but it is the information encoded on their surfaces that matters, information to be read off by the machine. Once the right software is combined with the appropriate hardware, one is in business. The program may then be run.

Life is very much like that. A living cell is made largely of proteins. This is the hardware. The membrane surrounding the cell is analogous to the plastic shell of my computer, or, perhaps more accurately, to the microchip substrate onto which the circuitry is etched. It's no good, however, just throwing a heap of proteins into a container and expecting life to happen. Even with the necessary raw materials, a cell won't do anything clever without software. This is normally provided by DNA. Like the floppy disk, DNA is itself hardware, but again the crucial feature is not the stuff of which DNA is made but the *message* written into its base pairs. Put this message into the right molecular environment—in the right *semantic context*—and, what do you know, life happens!

So life is a felicitous blend of hardware and software. More than mere complexity, it is informed or instructed complexity. Let me illustrate this subtle but absolutely crucial point with a couple of analogies. The nineteenth century was the great Age of the Machine. Many clever devices were invented. Take, for example, the steam-engine governor, a pair of balls attached to levers that rotate at a rate determined by the steam pressure. If the pressure gets too high, the balls whirl so fast that, by centrifugal force, they lever a valve open, thereby reducing the pressure. Today we would describe the principle behind this type of mechanism as "feedback." You wouldn't do it with balls any more. Instead, a sensor would feed data about the pressure electrically to a small computer or microprocessor. This electronic system would then process the information and instruct the valve to open or close using a motor. My wife's Holden Berina car has one of these microprocessors to maximize fuel efficiency. It decides how fast the engine should run when it is idling. The difference between the push-pull mechanical steam governor and the electronic microprocessor is that the former is a hardware solution to a problem and the latter depends on information processing and software, i.e., it is "digital."

The power of software is that it can act as an interface between chalk and cheese—different sorts of hardware that otherwise could not deal with each other effectively. Compare the difficulty of trying to steer a kite with the ease of flying a model aircraft by remote control. The difference here reduces to hardware versus software. The pull of the kite strings is a direct but very clumsy way of coupling the kite hardware to the control hardware (the person on the ground). The radio system, which first encodes the instructions and then relays the coded data to be interpreted at the other end, works much more efficiently. Of course, the informational flow from ground to aircraft may also be described in hardware terms: radio waves propagate from the transmitter to the receiver, where they induce an electric current that triggers circuits and moves airfoils, etc. However, this hardware description is merely incidental to the performance of the plane. The role of the radio waves is simply to serve as an information channel. The waves themselves don't push and pull the aircraft about. Instead, the coded information harnesses other, more powerful, forces to do the job.

A lumbering kite is a (literally) hard-wired mechanism, whereas the more efficient radio-controlled plane is an information-controlled mechanism. In a living organism we see the power of software, or information processing, refined to an incredible degree. Cells are not hard-wired, like kites. Rather, the information flow couples the chalk of nucleic acids to the cheese of proteins using the genetic code. Stored energy is then released and forces are harnessed to carry out the programmed instructions, as with the radio-controlled plane.

Viewed this way, the problem of the origin of life reduces to one of understanding how encoded software emerged spontaneously from hardware. How did it happen? How did nature "go digital"? We are dealing here not with a simple matter of refinement and adaptation, an amplification of complexity, or even the husbanding of information, but a *fundamental change of concept*. It is like trying to explain how a kite can evolve into a radio-controlled aircraft. Can the laws of nature as we presently comprehend them account for such a transition? I do not believe they can. To see why not, it is necessary to dig a bit deeper into the informational character of life.

A code within the code?

I have explained how life, at rock bottom, has the same logical structure as a computer. This fact provides an opportunity to inject some precision into the rather slippery notions of complexity and biological information by appealing to the theory of computation. (The reader need not despair; I shall not be resorting to advanced mathematics.) Much of the bafflement about life is due to confusion in the meaning of terms like "order," "organization," "entropy," "chance," "randomness," "information," and "complexity." These words are frequently employed in a slipshod or ambiguous way, without any proper definition. In particular, "order" and "organization" are often conflated.

First off, let's look at randomness. I shall take as an elementary example a string of ones and zeros. Figure 4.4 shows such a string. It is clearly not random, but periodic. A useful way to express the patterning shown is in terms of information. (The binary system of zero and one can, of course, be used to encode information; that is the way most ordinary computers do it.) We could abbreviate the entire information content of figure 4.4 to the simple statement "Print 10 twenty-five times." If I had chosen to fill the page with a continuation of this binary string, the abbreviated statement would scarcely be any longer. In other words, we can compress the information of a periodic sequence into a compact formula—or an algorithm, as mathematicians call it. A computer algorithm is just a recipe, or mechanical procedure, for generating some output. In the case under discussion, the elementary algorithm "Print 10 twenty-five times" generates the string shown in figure 4.4.

The reason we can compress the long string of digits into a few basic instructions is that the sequence has a regular pattern to it. We can imagine more complicated patterns too, that are still expressible by a comparatively short formula or algorithm. By contrast, if a string of zeros and ones had no pattern whatsoever—if it was random—then we would not be able to find a shortened description of it. No tidy little algorithm would be able to generate it as the output of a simple computational process. Gregory Chaitin, a computer scientist working for IBM, has produced a powerful and comprehensive theory of algorithmic information and complexity, and applied

1010101010101010101010101010101010101010101010101010

Figure 4.4. A binary sequence with a simple repeating pattern. This has very low information content, because its construction can be described by a simple procedure or computer algorithm.

it to many physical examples, including biological systems.[7] He proposes a definition of a random sequence as one that cannot be algorithmically compressed: the shortest description of a random sequence is simply the sequence itself.

Using this "algorithmic," or computer-program, definition of randomness, it is obvious that a random sequence is also an information-rich sequence, because the information content cannot be compacted into a simple formula. By contrast, a nonrandom pattern, like the periodic sequence shown in figure 4.4, contains very little information, because it can be abbreviated into a simple description ("10 twenty-five times"). If the whole point of a sequence is to encode information, as in a genome, patterns are bad news. Randomness is the way to go.

Figure 4.5 shows a string of ones and zeros that *looks* pretty random. But can we be sure? How do we know that there isn't a subtle pattern lurking in the sequence? Actually, there is. The sequence shown is the first fifty digits of the number pi, expressed in binary. It can be generated by a few lines of computer programming based on a simple formula. However, if you didn't know this you wouldn't spot any pattern: the sequence satisfies all the usual statistical tests for randomness. Yet pi is not random, using the algorithmic definition.

So far I have restricted the discussion to mathematics. How about nature? We can use the concept of algorithmic randomness to give a very powerful expression to the notion of a law. A law of nature is, in essence, just a simple way to describe (or predict) complicated behavior. To take a well-known example, consider eclipses of the Sun. If you wrote down the date of each successive eclipse and

110010010000111111011010101000100010000010110100011 . . .

Figure 4.5. Randomness? This binary sequence looks random, and has no known patterns, yet it contains hidden order. It is in fact the digits of pi, which can be generated by a simple algorithm. It is therefore not random after all, and in one sense the sequence contains little information.

expressed it in binary, you would get a string of ones and zeros that looked random. But that appearance would be deceptive. We can use Newton's laws to predict the dates of eclipses, and all other features of planetary orbits. Newton's laws are simple mathematical formulae that can be written on a small postcard, so the information about all those eclipses, and in fact the positions of the Earth and Moon on every day of the year, is already implicit in a rather short algorithm. The Earth-Sun-Moon system is therefore relatively information-poor, exhibiting many deep patterns and regularities.[8]

The patterning, or order, displayed in the motion of the planets, and represented by the existence of a simple Newtonian algorithm, is one example of a law of physics. Quite generally, when we say that a law is at work, we mean that the data describing the behavior of the system are nonrandom, and the future of the system can be accurately predicted with a rather simple formula.

We can now see the true nature of the biological mystery. Figure 4.6 shows yet another binary sequence. This time it is part of the genome of the virus MS2, expressed[9] using the assignments A = 00, U = 11, G = 01, C = 10. Now ask the question, is the sequence displayed in figure 4.6 random,[10] or does there exist a simple formula or algorithm that could generate it as the output of a computational process? In other words, is there a code within the genetic code—a palimpsest—that spells "organism"? Most people would, I think, answer no. They intuitively feel that the sequence should be random. Why? Well, suppose I were to exhibit the genome of a human being instead of MS2. It seems repugnant to suggest that our essential makeup, including much of our personality, could be "reduced to a mere formula." Surely there must be more to a human being (even a virus) than can be captured in some trivial handle-turning computation? Imagine if you were—body and soul, so to speak—little more than the square root of some undistinguished number, cranked out on a molecular machine using a four-letter alphabet!

There is also a less emotive reason for believing that the genome is mainly random. The job of a genome is, after all, to store genetic information. Given the complexity and almost limitless variability

... 01000111011101001001110011010110101110110101000010 ...

Figure 4.6. Random genome? This is a portion of the genome of the virus MS2. It must be (almost) random if it is to contain a lot of genetic information. The amino acids in the proteins it codes will therefore be linked in a random order.

of living things, so necessary to enable them to adapt, there must be a lot of specific information contained in each genome. But if genomes are information-rich, as is required for their biological function, then they have to be random (or almost so).[11] A periodic genome, for example, would be hobbled with a repetitive genetic message as useless as a stuck record. There is no code within the code.

Now, you might be thinking that, if biological organization is random, its genesis should be easy. It is, after all, a simple matter to create random patterns. Just take a jar of coffee beans and tip them on the floor. Surely nature is full of haphazard and chaotic processes that might create a random macromolecule like a genome?

This is a good question, and it marks the point where we encounter the truly subtle and mysterious nature of life in the starkest manner. Fact one: the vast majority of possible sequences in a nucleic-acid molecule are random sequences. Fact two: not all random sequences are potential genomes. Far from it. In fact, only a tiny, tiny fraction of all possible random sequences would be even remotely biologically functional. A functioning genome is a random sequence, but it is not just *any* random sequence. It belongs to a very, very special subset of random sequences—namely, those that encode biologically relevant information. All random sequences of the same length encode about the same *amount* of information, but the *quality* of that information is crucial: in the vast majority of cases it would be, biologically speaking, complete gobbledygook.

The conclusion we have reached is clear and it is profound. A functional genome is *both* random *and* highly specific—properties that seem almost contradictory. It must be random to contain substantial amounts of information, and it must be specific for that information to be biologically relevant. The puzzle we are then faced with is how such a structure came into existence. We know that chance can produce randomness, and we know that law can produce a specific, predictable end-product. But how can both properties be combined into one process? How can a blend of chance and law cooperate to yield a *specific* random structure?

To get some idea of what we are up against with this dilemma, it is rather like tipping out the coffee beans from a jar to make a *particular* random pattern. Not just any old random pattern, but a definite, narrowly specific, predetermined random pattern. The task seems formidable. Could a law on its own, without a huge element of luck (i.e., chance), do such a thing? Can *specific randomness* be the guaranteed product of a deterministic, mechanical, lawlike process, like a primordial soup left to the mercy of familiar laws of physics and chemistry? No, it couldn't. No known law of nature could achieve this—a fact of the deepest significance, as we shall see in the final chapter.

If you have found the foregoing argument persuasive, you could be forgiven for concluding that a genome really is a miraculous object. However, most of the problems I have outlined above apply with equal force to the evolution of the genome over time. In this case we have a ready-made solution to the puzzle, called Darwinism. Random mutations plus natural selection are one surefire way to generate biological information, extending a short random genome over time into a long random genome. Chance in the guise of mutations and law in the guise of selection form just the right combination of randomness and order needed to create "the impossible object." The necessary information comes, as we have seen, from the environment.

Now, Darwinian evolution is a long and arduous process. Life has to struggle very hard to elaborate its gene pool this way. So what about the first genome? Was it too the product of an equally toughgoing evolutionary process, or did its complexity come for free? Computer scientists know of certain computational problems that are *irreducibly* complex: that is, they cannot be reduced to simple, nifty procedures. A famous example is the so-called traveling-salesman problem, which involves working out the shortest route that a salesman should take through a collection of cities without visiting each more than once. Problems like this turn out to be computationally intractable, not because they cannot be solved, but because the amount of computation required escalates with their size (with the total number of cities in the example cited).

It appears as if the information processing needed to generate a genome might also be computationally intractable. Sorting out a particular random sequence from all possible sequences looks like a problem every bit as daunting as that of a traveling salesman faced with visiting a million cities. Which casts the central paradox of biogenesis in the following terms. Given that it requires a long and arduous computation (i.e., a sequence of information-processing steps) to evolve a genome from microbe to man, could the (already considerable) genome of a microbe come into being without a comparably long and arduous process? How, in the phase before Darwinian evolution kicked in, could a *very particular* sort of information have been scavenged from the nonliving environment and deposited in something like a genome?

Viewed in the light of the theory of computation, the problem of biogenesis appears just as perplexing as it does through the eyes of the physicist or chemist. And the difficulties are not purely technical. Thorny philosophical problems loom too. Concepts like information and software do not come from the natural sciences at all, but from communication theory (see chapter 2), and involve qualifiers like context and mode of description—notions that are quite alien to the physicist's description of the world. Yet most scientists accept that informational concepts do legitimately apply to biological systems, and they cheerfully treat semantic information as if it were a natural quantity like energy. Unfortunately, "meaning" sounds perilously close to purpose, an utterly taboo subject in biology. So we are left with the contradiction that we need to apply concepts derived from purposeful human activities (communication, meaning, context, semantics) to biological processes that certainly appear purposeful, but are in fact not (or are not supposed to be).

There is clearly a danger in science of projecting onto nature categories and concepts derived from the world of human affairs as if they are intrinsic to nature itself. Yet, at the end of the day, human beings are products of nature, and if humans have purposes, then at some level purposefulness must arise from nature and therefore be inherent in nature. Is purposefulness a property that emerges only at the relatively high level of *Homo sapiens*, or does it exist in other animals too? When a dog seeks out and digs up a buried bone, does it "desire" to retrieve the bone? When an amoeba approaches and engulfs a particle of food, does it in any sense "intend" to swallow it? Might purpose be a genuine property of nature right down to the cellular or even the subcellular level? There are no agreed answers to these questions,[12] but no account of the origin of life can be complete without addressing them.

CHAPTER 5

The Chicken-and-Egg Paradox

SOME YEARS AGO, BBC television screened an engaging, if some-what depressing, science-fiction series called *The Survivors*, about a disease that wiped out most of humanity. Just a handful of people remained alive, forced to eke out an existence as best they could. Reduced to scavenging, this beleaguered community soon exhausted its resources and was faced with extinction. In a fit of pessimism, the two central characters began to squabble. What would happen when even basic commodities ran out? the woman asked. Her partner put a brave face on it: people would just have to start making things for themselves. Give him a saw, he suggested reassuringly, and he would be quite capable of making a table. "But what happens when the last saw breaks?" the woman retorted. "You haven't got the tools to make the tools."

The predicament of the survivors illustrates just how dependent we are on each other in our modern technological society. Everybody needs everybody else to keep the whole thing going. As such, the story is a metaphor for all life. The cell is an elaborate self-sustaining community of molecules, each dependent on the others.

Take DNA. Despite its much-vaunted longevity, it can't do a lot on its own, because it is chemically impotent. It has a grand agenda, but to implement this, DNA must enlist the help of proteins. As I have explained, proteins are made by complicated machines called ribosomes, according to coded instructions received from DNA via mRNA. The problem is, how could proteins get made without the DNA to code for them, the mRNA to transcribe the instructions, and the ribosomes to assemble them? But if the proteins are not already there, how can DNA, ribosomes and all the rest of the paraphernalia get made in the first place? It's Catch-22.

All known life revolves around the cozy accommodation between DNA and proteins: the software and the hardware. Each needs the other. So which came first? We have already encountered this sort of chicken-and-egg paradox in chapter 2, concerning the so-called error catastrophe that limits the number of copying mistakes in genetic replication, but the problem is much more general. There seems to be an enigmatic circularity to life, a type of irreducible complexity that some people regard as utterly mysterious.[1]

In this chapter I shall discuss various attempts to break out of this vicious circle, but first let me make a general point. The BBC drama reminds us that complex systems can get themselves irreversibly into cycles of dependence. Nobody suggests that technological society cannot have arisen by gradual evolution just because we all need each other today. Take a simple case. A blacksmith shapes iron using iron tools: he needs iron instruments to make iron instruments. So where did the first iron instruments come from? Must they have been handed down from on high, ready-made? Of course not. Early blacksmiths might have used stone clubs, for example, or other metal tools, to shape the first iron hammers.

It is possible to evolve sophisticated technological cycles in many different ways from fumbling beginnings, but once a cycle is established it rapidly becomes refined. When that happens, few traces of its low-tech origins survive. Today's organisms are full of high-tech chemical cycles that must somehow have emerged from long-discarded molecular groping. We can glimpse a general principle that helps explain how this may have happened. If A needs B and B needs A, there is a type of causal-feedback loop. A small change in A has repercussions for B, which in turn have an impact on A, and so on, round and round the loop. Causal feedback can produce dramatic amplification effects. If an accidental improvement occurs in, say, A in such a way as to improve B, which in turn improves A, then the improvement is rapidly reinforced.

Nobody expects that nucleic acids and proteins came into being ready-made, with their mutually beneficial properties already inscribed. A cruder association of chemicals must have arisen first, to be honed into its present form by a succession of feedback loops combined with Darwinian selection. Somehow, along the way, a separation occurred into hardware and software, chicken and egg. That much is agreed. But where agreement fails is in the basic order of events: controversy rages over what started it all off.

RNA *first*

A glance at the chain of command in the modern cell reveals DNA as the boss, running the show by its coded instructions, getting RNA to do all the fetching-and-carrying work, and telling the ribosomes what proteins to make next. The proteins have a completely servile role, but are the real workers.

As I have explained, DNA is a chemical dud, but its first cousin RNA is rather more potent. RNA is in fact very versatile, carrying out several essential tasks in the cell that look as if they date back to the earliest life forms. Among its many functions, RNA transcribes and relays the instructions from DNA. It therefore plays a crucial but subordinate role genetically. Nevertheless, RNA is built of (almost) the same four-letter alphabet as DNA, and it could act as a genome. As a matter of fact, it does sometimes act as a genome: certain viruses use RNA in place of DNA. So RNA can certainly do the job of genetic storage. It is more fragile than DNA, but by no means useless.

In the 1960s, Leslie Orgel of the Salk Institute in La Jolla, California, suggested that maybe RNA came first, not just before DNA, but before proteins too. The obvious question is, what played the role of enzymes in the absence of proteins? A possible answer came in 1983. Thomas Cech and co-workers at the University of Colorado, and Sidney Altman and his team at Yale, discovered that RNA is chemically active enough to behave as a weak catalyst itself.[2] Although it cannot match the catalytic prowess of proteins, RNA can mimic certain enzymes that facilitate the cleavage and joining of other RNA strands. Biochemists were quick to spot that if RNA could somehow catalyze its own replication then life may have begun with a soup of RNA molecules acting *both* as genetic storehouses and, when folded into suitable three-dimensional shapes, as catalysts. Hardware and software would be present in the same group of molecules.[3] This theory became known as the RNA world.

Advocates of the RNA-world theory suppose that a soup containing RNA molecules can evolve by a type of Darwinian process. Normally, Darwinism is associated with organisms like cells, but in principle all it needs is replication, variation, and selection. These can occur even at the molecular level, and biochemists use the terms "molecular evolution" or "molecular Darwinism" to describe what happens. It is a moot point whether we should define as living anything that can evolve in a Darwinian manner. If so, then perhaps RNA molecules (in a suitable chemical environment) could already be considered as living things.

A famous experiment carried out in the late 1960s attempted to demonstrate how Darwinism might act at the molecular level.[4] It was based on a small RNA virus named Q_B. A virus is simply a strand of DNA or RNA encased in a protein coat. Although viruses store genetic information, they cannot replicate on their own. To do so, they invade cells and hijack their reproductive apparatus, adapting it to make more viruses. That some viruses use RNA for a genome implies that they might be surviving relics of an RNA world.

The Q_β virus doesn't need anything as complicated as a cell in order to replicate: a test tube full of suitable chemicals is enough. The experiment, conducted by Sol Spiegelman of the University of Illinois, consisted of introducing the viral RNA into a medium containing the RNA's own replication enzyme, plus a supply of raw materials and some salts, and incubating the mixture. When Spiegelman did this, the system obligingly replicated the strands of naked RNA. Spiegelman then extracted some of the freshly synthesized RNA, put it in a separate nutrient solution, and let it multiply. He then decanted some of *that* RNA into yet another solution, and so on, in a series of steps.

The effect of allowing unrestricted replication was that the RNA that multiplied fastest won out, and got passed on to the "next generation" in the series. The decanting operation therefore replaced, in a highly accelerated way, the basic competitive process of Darwinian evolution, acting directly on the RNA. In this respect it resembled an RNA world.

Spiegelman's results were spectacular. As anticipated, copying errors occurred during replication. Relieved of the responsibility of working for a living and the need to manufacture protein coats, the spoon-fed RNA strands began to slim down, shedding parts of the genome that were no longer required and merely proved to be an encumbrance. The RNA molecules that could replicate the fastest simply out-multiplied the competition. After seventy-four generations, what started out as an RNA strand with 4,500 nucleotide bases ended up as a dwarf genome with only 220 bases. This raw replicator with no frills attached could replicate very fast. It was dubbed Spiegelman's monster.

Incredible though Spiegelman's results were, an even bigger surprise lay in store. In 1974, Manfred Eigen and his colleagues also experimented with a chemical broth containing Q_β replication enzyme and salts, and an energized form of the four bases that make up the building blocks of RNA.[5] They tried varying the quantity of viral RNA initially added to the mixture. As the amount of input RNA was progressively reduced, the experimenters found that, with little competition, it enjoyed untrammeled exponential growth. Even a single RNA molecule added to the broth was enough to trigger a population explosion. But then something truly amazing was discovered. Replicating strands of RNA were still produced even when *not a single molecule* of viral RNA was added! To return to my architectural analogy, it was rather like throwing a pile of bricks into a giant mixer and producing, if not a house, then at least a garage. At first Eigen found the results hard to believe, and checked to see whether accidental contamination had occurred. Soon the experimenters convinced themselves that they were witnessing for the first time the spontaneous synthesis of RNA strands from their basic building blocks. Analysis revealed that under some experimental conditions the created RNA resembled Spiegelman's monster.

To some observers, Eigen's experiments already amount to the creation of life in a test tube. Spiegelman, remember, extracted RNA from a virus that by some definitions would be considered a living thing. Following a continuous sequence of steps, he then produced a test-tube mutant that was much smaller but still capable of replication. Eigen started from the bottom up, achieving molecular self-assembly from simple building blocks, and met Spiegelman partway by producing a replicating RNA molecule similar to Spiegelman's once-living derivative. No dividing line crossed this path; nothing separated the realm of the living from the nonliving. A sequence had been identified that led seamlessly from a simple chemical mixture up through a viable virus.

Do Eigen's experiments re-create the steps that nature took in making life from nonliving materials? Clearly not. Exciting though the experiments may be, they are highly contrived, and a world away from the natural conditions that prevailed on the young Earth. In particular, to achieve RNA synthesis, Eigen had to use a very carefully prepared chemical mixture that, crucially, included a customized replication enzyme that was extracted from a living organism. This enzyme is highly specialized, and is not the sort of molecule that would have been lying around on Earth prior to life. Eigen is a long way from demonstrating that nucleic-acid bases will spontaneously assemble and replicate in an incoherent mixture like a primordial soup.

Many biochemists concede this, and question whether RNA was in fact the first replicating molecule on the block. After all, template replication might work with lots of other sorts of molecules, including simpler and more easily synthesized structures. Once template replication got under way, it could have been successively refined by molecular evolution. Each mutation that increased the efficiency of the replication process would spread rapidly through the chemical soup by the multiplier effect. At some stage, this steady refinement process might have produced RNA as the best replicator around. Possibly the first RNA molecules contained additional bases, not just the four used today. However, the snug two-by-two complementarity of the four surviving nucleotides ensured that they would eventually be chosen and the others discarded in the replication game. During this period of prebiotic groping, replication would have been highly inefficient by present standards, because the broth lacked the all-important enzymes needed to make the process whizz along.

If we accept this scenario for now, the next question to address is how a limited RNA world evolved into the present dual system of nucleic acids and proteins linked by a genetic code. Researchers surmise that the primitive gene was a precursor of the modern transfer RNA. They cite two reasons for homing in on this molecule. First, tRNA has evolved very little over time: some human and frog tRNA molecules are identical. This suggests that tRNA has a long history. Second, the very job of tRNA is to link up with appropriate amino acids, the stuff of proteins. So the glimmering of an association is there. Amino acids would undoubtedly have been plentiful in any primordial soup. An RNA molecule that could interact with amino acids had the potential to join them up into proteins. The next step in the RNA world would have been for these primitive RNA strands to start making proteins accidentally. Nobody knows how this pivotal event may have come about; there are theories but few hard facts. It may have begun with nothing more sensational than two RNA molecules colliding, and one transferring its cargo to the other to make a double amino-acid strand hanging from the RNA. Then a third amino acid could have been added, and so on. Some of these primitive polypeptides would doubtless have had a favorable effect on the RNA replication, and so a self-reinforcing cycle would be established: RNA made proteins, which in turn accelerated the production of more RNA and more proteins, and so on. The proteins that most effectively aided RNA replication would be rewarded with more copies of themselves. In this way, step by convoluted step, the intimate partnership of nucleic acids and proteins would become established. At least, that's the story.

A tricky problem that might be solved by this theory is how to circumvent the error-catastrophe trap (see page 59). Recall that long RNA chains are most vulnerable to copying errors, but short ones can't store enough information to make good copying machinery. However, a collection of several short RNA molecules might cooperate and share the genetic burden between them. Imagine a closed cycle of chemical reactions in which several RNAs catalyze each other's replication: for example, A makes B, B makes C, C makes D, and D makes A. The system thus forms a self-reinforcing reaction loop, termed a hypercycle. If such a chemical loop becomes enclosed in a membrane, in the fashion of a primitive cell, it has the possibility of mutating in a way that improves the efficiency of the replication process. If the cell also divides by simple mechanical fission, this successful mix of chemicals may be inherited by the daughter cells. In this manner a rudimentary type of evolution may be possible as the cells containing the more efficient hypercycles out-replicate the others.[6]

Promising though the RNA-world scenario seems, it has many detractors. They point out that, however good the theory may be, test-tube experiments are frequently dismal failures. Key reactions stubbornly refuse to proceed without carefully designed procedures and the help of special catalysts. Nucleic-acid chains are notoriously fragile, and tend to snap long before they have acquired the fifty or so base pairs needed for them to act as enzymes. Water attacks and breaks up nucleic-acid polymers as it does peptides, casting doubt on any soupy version of an RNA world. Even the synthesis of the four bases required as building blocks is not without serious problems. As far as biochemists can see, it is a long and difficult road to produce efficient RNA replicators from scratch. No doubt a way could eventually be found for each step in the chemical sequence to be carried out in the lab without too much drama, but only under highly artificial conditions, using specially prepared and purified chemicals in just the right proportions. The trouble is, there are very many such steps involved, and each requires different special conditions. It is highly doubtful that all these steps would obligingly happen one after the other "in the wild," where a chemical soup or scum would just have to take pot luck.

The conclusion has to be that, without a trained organic chemist on hand to supervise, nature would be struggling to make RNA from a dilute soup under any plausible prebiotic conditions. So, although an RNA world could conceivably function and evolve towards life if handed to us on a plate (perhaps in a soup bowl would be a better metaphor), getting the RNA world going from a crude chemical mixture is another matter entirely.

Added to these diverse difficulties is the problem of chirality—left versus right—that I mentioned in chapter 3. That all life on Earth is based on molecules with the same chirality is not merely a curiosity: RNA replication would be menaced in an environment in which both left- and right-handed versions of the basic molecules were equally present. The crucial lock-and-key templating arrangements, whereby bases pair up with complementary bases according to their shapes, would be compromised as molecules with the "wrong" handedness locked into the slots. The left hand would mess up what the right hand was doing. Unless a way could be found for nature to create a soup with molecules of only one handedness, spontaneous RNA synthesis would be a lost cause.

Proponents of the RNA-world scenario have received flak not just from chemists but from biologists too. If life began with RNA replication, you would expect the necessary replication machinery to be very ancient, and therefore common to all extant life. However, genetic analysis reveals that the genes coding for RNA replication differ markedly in the three domains of life, suggesting that RNA replication was refined sometime *after* the common ancestor lived.

There has also been criticism on theoretical grounds. The RNA-world theory focuses exclusively on replication at the expense of metabolism. As I have stressed already, life is about more than raw reproduction: living organisms also do things, and must do them if they are to survive to reproduce. Doing things costs energy. There has to be a ready source of energy for organisms to metabolize. In test-tube experiments, RNA molecules are lovingly supplied with specialized energetic chemicals to power their activities; in nature, RNA would have to make do with whatever was lying around. No Miller-Urey type of experiment has succeeded in fabricating the energizing chemicals used by extant life: they are all manufactured inside cells. Spoon-fed RNA may be a slick replicator, but without an energy-liberating metabolic cycle already in place, these fecund molecular strands would soon become genetic dropouts.

An obvious escape route is to seek a self-replicating molecule far simpler than RNA to start the whole game going. The RNA world would then come only much later. It is conceivable that a relatively small molecule might be found that could replicate faithfully enough. The way would then lie open for molecular evolution to elaborate it, adding information step by step, until a level of complexity comparable to short strands of RNA was achieved. The system could then be "taken over" by RNA.[7]

Is this how biogenesis really happened? Maybe. However, there are many obstacles to the theory, such as doubt over whether small molecules can be accurate enough replicators to avoid the error catastrophe. In extant life, high-fidelity replication seems to be associated with large, complex systems. The larger genomes, with their editing and error-correcting procedures, are the best copiers. So, if the trend among nucleic-acid replicators is followed down to smaller and smaller size, one expects only poor replication accuracy from simple molecules. Moreover, the smaller a molecule is, the more drastic will be the relative effect of any mutational change, and the greater the chance that the mutation won't inherit the property of being a replicator itself.

In recent years, attempts have been made to build small and simple replicator molecules in the lab, and to subject them to environmental stresses to see if they evolve into better replicators.[8] Modest success has been claimed. However, these experiments do not demonstrate molecular evolution in nature. They have yet to show that the sort of small replicators that have been painstakingly designed and fabricated in the laboratory will form spontaneously under plausible prebiotic conditions, and if they do, whether they will replicate well enough to evade the error catastrophe. In short, nobody has a clue whether naturally occurring mini-replicators are even possible, let alone whether they have got what it takes to evolve successfully.

RNA last

A completely different way to solve the chicken-and-egg paradox is to invert the order of events and assume that proteins came first and nucleic acids came afterwards. The big problem is then to understand how proteins can replicate without nucleic acid to relay the necessary instructions. Can proteins replicate unaided? Recently, Reza Ghadiri of the Scripps Institute in San Diego discovered that some small peptide chains can indeed self-replicate. Moreover, they can apparently correct replication errors "as if they had a mind of their own."[9] Another clue comes from the infamous mad-cow disease, or BSE, which has decimated British cattle stocks. Like scrapie and kuru, BSE is caused not by a bacterium or a virus, but by a fragment of protein that can replicate and spread. Might such fragments be surviving relics of a primitive life form based solely on proteins?

The most distinguished proponent of the proteins-first theory is Freeman Dyson, a now retired physicist from Princeton's Institute for Advanced Study. Dyson argues that life really had two origins: one for the hardware, another for the software.[10] He envisages two varieties of primordial creature, one capable of protein metabolism but unable to replicate properly, and another that could replicate but had no metabolism. Life as we know it arose from a fusion, or symbiosis, of the two. Dyson takes his cue from Oparin and his followers, who maintain that the very first step towards life involved the formation of some sort of cells or vesicles. We can think of these proto-cells as naturally occurring test tubes containing concentrated primordial soup.

Because they lack a genome, Darwinian evolution isn't an option for Dyson's cells, but they might still evolve by chemical means. To investigate how, Dyson formulated a mathematical model to describe a chemical mixture, such as a soup of amino acids that changes with time as the chemicals react in complicated ways. Especially important in Dyson's model is the assumption that molecules can catalyze the production and mutation of other molecules. The upshot of this mathematics is the prediction of spontaneous transitions from disorder to order. Here disorder means a chaotic assemblage of molecules, and order means certain preferred chemical cycles, reminiscent of metabolism. Dyson's chemical bags are not replicators; their order arises spontaneously, rather than by genetic specification. The production of molecules within the cells is therefore very imprecise.

Although Darwinian evolution needs some form of heritable replication plus natural selection, it is possible to conceive of other, weaker, forms of selection that might serve to produce a rudimentary kind of evolution, to get the thing started. Once there exists a growing population of distinct cells, even if they are just globs of chemicals that swell and fission, then a type of competition is inevitable. Some cells will grow and split faster than others because of their "better" internal chemistry and come to outnumber their competitors. If the cells can pass on at least some of their chemical characteristics, and if resources are limited, the most "successful" cells (from a chemical point of view) will prevail. The challenge is then to explain how this rather hit-and-miss selection turned into the more precise gene-based natural selection of conventional Darwinism.

A possible solution is parasitism. Dyson suggests that the geneless cells were invaded by primitive nucleic-acid replicators, and the two systems melded. The nucleic-acid parasites found that the bags of proteins aided their replication process. Obviously it would have proved advantageous for the replicators also to replicate the helpful proteins along the way, to boost their own replication. Given the cellular structure, natural selection would cut in at this point, pitting cell against cell, escalating the rate of evolutionary improvements. Selection would strongly favor replicators that made some or all of the necessary ingredients of the protein cells, and a full symbiosis would rapidly emerge, leading to life as we know it.

Where might all this have taken place? Oparin envisaged his coacervate cells in some pond or sea, but if life started on or beneath the seabed, as recent evidence suggests, then oily blobs may not be the answer. The porous basalt rock of the sea floor provides a natural network of tiny tunnels and cavities which could trap large organic molecules. The mineral surfaces might also act as convenient catalysts and serve to concentrate the organic material. Unfortunately, rock cavities can't multiply by fission. Euan Nisbet of the University of London has suggested that perhaps membranes might form within cavities, like creatures trapped in tiny caves, to be liberated in due course by some geological upheaval.[11]

Another imaginative idea for a primitive cell has been proposed by Mike Russell of the University of Glasgow.[12] His theory focuses on regions of the seabed somewhat removed from volcanic vents, where water seeps gradually into the rock to a depth of several kilometers. Convection eventually returns it to the surface, rich with dissolved minerals. The emerging water is alkaline, and very hot—perhaps reaching two hundred degrees Celsius under high-pressure conditions. By contrast, the overlying ocean would have been acidic, on account of dissolved carbon dioxide, and much cooler. Russell has found that the conjunction of the two fluids triggers the formation of a colloidal membrane made of iron sulfide. As we shall see, iron and sulfur are two chemicals strongly implicated in early life. Moreover, the membrane is semipermeable: it lets through some chemicals but not others, just like a living cell. Russell has managed to grow large cell-like bubbles in the laboratory, and has found evidence for similar structures fossilized in Irish rocks. He believes that osmotic and hydraulic pressure would inflate the bubbles and make them divide. A bonus of his theory is that the juxtaposition of acid-membrane-fluid acts like an electrical battery, which could have provided the initial power source to drive early metabolism. In modern cells there is also a small voltage across the membrane. So maybe electricity was, after all, the original life force!

A completely different theory for the origin of life has been given by the British biochemist Graham Cairns-Smith, also from the University of Glasgow, who shares the belief that nucleic acids came late in the piece.[13] In fact, as far as the chicken-and-egg (or nucleic-acids-and-proteins) argument goes, he thinks that life started with neither. Cairns-Smith begins by reminding us that nucleic acids function primarily as software—the repositories of genetic information. That being so, their specific chemical form is irrelevant. Just as we can store the same digital information on magnetic tape or floppy disk, so genetic information could be contained in physical structures other than RNA or DNA. Perhaps life started with information encoded in some other manner, and only at a relatively late stage was the genetic function entrusted to nucleic acids.

What sort of structures might serve to store the original genetic database? Cairns-Smith suggests that clay crystals offer an attractive possibility. Less fragile than nucleic acid, crystals can nevertheless replicate after a fashion. Clay particles can be infused with metallic ions in an irregular way, and information could in principle be encoded in their patterns, to be reproduced as the crystal grows, layer by layer. Dirty crystals may not strike the reader as very lifelike, but the essential properties necessary for evolution—replication, variation, and selection—could all have been manifested in clay.

Once crystal evolution got under way, the stage would be set for the next step: organic molecules. Perhaps these were initially manufactured by the clay crystals for their own ends, such as speeding replication, cementing crystal faces, or any number of secondary tasks. Whatever they were used for, they would need to confer some selective advantage for evolution to refine them. Any crystals that discovered how to make self-replicating nucleic acids would be onto a winner, for they would then have a ready supply of these presumably handy substances available. But in crossing this line, crystal life would have sown the seeds of its own demise. Once nucleic acids began to out-replicate their crystalline creators, they would rapidly take over and become the prevailing life form. The poor lumbering crystals would be rendered obsolete.

It has to be said that there is very little experimental evidence to support Cairns-Smith's clay theory. Still, whatever the plausibility of clay as the primal life stuff, the basic principle of genetic takeover is sound. Everybody agrees that the existing system of nucleic acids and proteins is too complex to arise in one go as a ready-made system. Even though all extant life is based on nucleic acids and proteins, life didn't have to begin that way. If there is a simpler route from nonlife to life, then the present biochemical arrangements could be the refined derivative of a low-tech precursor.

Cairns-Smith uses the analogy of a stone arch to illustrate the transition from "low-tech" to "higher-tech" life. An arch seems baffling at first sight. The assembled structure is self-supporting, but half an arch would collapse. How, then, did the arch come to be? The answer is, a scaffold was used to build it. So we should look for molecular scaffolding that may have been used to build nucleic acid. Perhaps clay crystals are part of the answer, or perhaps it needs some other system we haven't thought of yet. But whatever it was, once RNA life got going, the scaffold was discarded and has long since been obliterated.

So what can be concluded from these various speculations about life's origin? They all share one assumption. Once life of some sort had established itself, the rest was plain sailing, because Darwinian evolution could then take over. It is therefore natural that scientists should seek to invoke Darwinism at the earliest moment in the history of life. As soon as it kicks in, dramatic advances can occur with nothing fancier than chance and selection as a driving force. Unfortunately, before Darwinian evolution can start, a certain minimum level of complexity is required. But how was this initial complexity achieved? When pressed, most scientists wring their hands and mutter the incantation "Chance." So, did chance alone create the first self-replicating molecule? Or was there more to it than that?

Self-organization: something for nothing?

Life is but one example of complexity found in nature. Many other examples occur in the world about us. We see complexity in the spangled pattern of frost on a window, in the intricate wiggles of a coastline, in the filigrees and whorls that adorn the surface of Jupiter, and among the jostling eddies of a turbulent river. Life is not haphazard complexity, it is organized. Disorganized complexity is found all over the place, from the spatter of raindrops on the ground to the tea leaves at the bottom of the cup. But organized complexity, though scarcer, is by no means restricted to biology. A spiral galaxy, a rainbow, and a diffraction pattern from a laser beam are both complex and organized. Yet they form without any genes to specify them or any Darwinian evolution to create them. If nonliving systems can generate organized complexity spontaneously, just by following the laws of physics, why can't life do it that way, at least in the beginning?

Some people think it can. The Belgian chemist Ilya Prigogine has given examples of chemical mixtures that behave in a lifelike manner, forming elaborate spirals or undergoing rhythmic pulsations.[14] The hallmark of these reactions is that they take place far from thermodynamic equilibrium, and require a continual throughput of matter and energy—as does life. The spontaneous ordering doesn't clash with the second law of thermodynamics because the systems are open; entropy is exported into the environment to pay for the increase in order. Characteristic of such self-organizing systems is their tendency to reach critical "bifurcation" or indecision points, where their behavior is unpredictable. They may leap abruptly to a new state of greater complexity and stabilize, or descend into chaos. Prigogine and his many devotees envisage a sequence of self-organizing transitions, where matter driven by an energy flow jumps to higher and higher levels of organized complexity, until it is truly living.

A simple and instructive example of self-organization is the formation of convection cells. If a pan of water is heated on a stove, the fluid near the bottom gets hotter than the fluid at the top. For gentle heating the water remains featureless: heat flows steadily upwards by conduction. Now consider what happens when you turn up the gas. The hot layer of water at the base wants to rise (being less dense), but is inhibited by the weight of the cooler layers above. Eventually the hot water breaks out in a rising plume, and convective motion begins. If the heating is done carefully, the convention pattern arranges itself into an ordered honeycomb of hexagonal cells. This stable configuration involves the cooperation of countless water molecules to produce large-scale order. The sudden transition to convective flow occurs when the system is forced far from thermodynamic equilibrium, and the resulting order is paid for by a flux of entropy from the pan into the surroundings. Without the gas to provide a source of free energy (i.e., to maintain a thermodynamic disequilibrium between the bottom and the top of the fluid), the convection cells would vanish and the state of the water would soon sag back to featureless equilibrium.

Stuart Kauffman, a biophysicist at the Santa Fe Institute for the Study of Complexity, has tried to flesh out the details of the self-organization route to life, focusing on a chemical phenomenon known as autocatalysis.[15] A catalyst, remember, is a type of molecule that promotes a reaction among other molecules without getting altered itself. Imagine, then, a primordial soup in which many different reactions are taking place together. Complex organic molecules are being created and destroyed, combining with other molecules, and splitting into fragments. There is a vast and elaborate network of reactions going on—a chemical ecosystem, if you like.

Now imagine that, in this seething stew, some molecules find themselves playing a dual role: on the one hand, they enter into certain chemical reactions as inputs or outputs; on the other hand, they also act as catalysts for other reactions. It may then happen that the presence of a particular molecule M has the effect of catalyzing the very reaction sequence that leads to the production of M itself. So the existence of M accelerates the creation of more M: hence the designation *auto*catalysis. When this process occurs, a feedback cycle is set up that grows stronger and stronger, forming a self-reinforcing web of reactions.

What happens next? When the diversity of molecules in the network is great enough, the system crosses a critical threshold. Kauffman predicts an abrupt leap into a gigantic autocatalytic cycle, a process of self-organization akin to the sudden transition from a featureless fluid to convection cells. This elevated and much more complex cycle will be a crude form of metabolism, the type of organized chemical processes that Oparin and Dyson envisioned for the contents of their chemical vesicles. No special molecule such as RNA is involved, and no genetic apparatus is needed. All that comes later.

Though autocatalytic cycles may seem complicated and contrived, they represent an example of a very widespread phenomenon. Computer models show that any network with enough components and interactions will tend to flip spontaneously into a state of organized complexity. Physicists see this phenomenon at work in magnetic materials, and economists see it in world financial markets. If Kauffman's ideas are on the right track, it may be that life is a consequence, not of special organic chemistry, but of universal mathematical rules that govern the behavior of all complex systems, regardless of what they are made of.

Attractive though self-organization may seem, it faces two major obstacles when it comes to the origin of life. The first is the paucity of convincing experiments. So far, most of the "experiments" have been computer simulations rather than the real thing. This has earned the subject of complexity theory something of a bad name in biology. In a now famous put-down of Kauffman's ideas, John Maynard Smith once described them, somewhat harshly, as "fact-free science."[16]

There is, however, a deeper problem of a conceptual nature. Life is actually *not* an example of *self*-organization. Life is in fact *specified*—i.e., genetically directed—organization. Living things are instructed by the genetic software encoded in their DNA (or RNA). Convection cells form spontaneously by self-organization; there is no gene for a convection cell. The source of order here is not encoded in software; it can instead be traced to the boundary conditions on the fluid. The flux of heat and entropy across the boundaries triggers the self-organization, and the shape, size, and nature of the boundaries determine the patterning details of the cells. In other words, a convection cell's order is imposed *externally*, from the system's environment. By contrast, the order of a living cell derives from *internal* control, from its genes, which are located on a microscopic molecule buried deep within the system that chemically broadcasts its instructions outwards. To be sure, the environment enveloping a living cell's membrane will influence to some extent what goes on within the cell, but the principal characteristics of an organism are determined by its genes.

The theory of self-organization as yet gives no clue how the transition is to be made between spontaneous, or self-induced, organization—which in even the most elaborate nonbiological examples still involves relatively simple structures—and the highly complex, information-based, genetic organization of living things. An explanation of this genetic takeover must account for more than merely the origin of nucleic acids and their potent entanglement with proteins at some later stage. It is not enough to know how these giant molecules arose or started to interact. We also need to know how the system's software came into existence. Indeed, we need to know how the very concept of software control was discovered by nature. To revisit the analogies I gave in chapter 4, we seek an explanation for how a kite can turn into a radio-controlled plane, or a steam-engine governor can evolve into a digital data-processing electronic regulator. This is not merely a matter of adding an extra layer of complexity; it is about a fundamental transformation in the very nature of the system.

Related to the latter criticism is the need to draw a careful distinction between order and organization. In the foregoing I have used the terms interchangeably, but they often have opposite meanings. Properly speaking, order refers to simple patterns. A periodic sequence of ones and zeros—like figure 4.4 on page 116, for example—is ordered. Likewise, a crystal is ordered. Both are highly non-random and so, as I explained in the last chapter, they cannot possess the complex organization and information storage of a genome. Attempts to seek a route to life via self-organization often fall into the trap of mistaking organization with order. Cited examples of self-organization are often nothing of the sort; rather, they involve spontaneous ordering instead. For instance, chemical reactions that display rhythmic cycles are often given in accounts of self-organization,[17] but periodic behavior is clearly a case of *nonrandom* order. Similarly, the hexagonal convection cells I described above are more reminiscent of crystalline order than of the organized complexity of biological organisms. In the absence of some new principle of self-organization that induces the production of algorithmic complexity, a crucial part of the biogenesis story has been left out.

So much for the bottom-up approach to the origin of life. It has yielded some useful pointers, but it leaves many bewildering riddles. However, it is not the only approach available. We can also pursue a top-down route. The idea here is to start with extant life and follow it back in time, hoping to guess where and how the earliest organisms lived. We can then employ this knowledge to tell us something about how these organisms may have come to exist. It turns out that, to track down the first living things on Earth, we must first take a look into space.

CHAPTER 6

The Cosmic Connection

SOME TWO HUNDRED KILOMETERS WEST of the town of Port Augusta in South Australia, in the rough outback country on the edge of the Nullarbor Plain, lies a large dried-up lake. Approximately circular in shape, Lake Acraman stretches thirty kilometers from side to side. Though it resembles many other salt basins in that part of Australia, Acraman is no ordinary lake bed. About six hundred million years ago, a giant meteor plunged from the sky and blasted an enormous hole in what is now the Eyre Peninsula. The original measured at least ninety kilometers across and several kilometers deep. Today's Lake Acraman is all that remains of this monstrous scar, a mute witness to an ancient cataclysm of impressive proportions.

The physical damage caused by a large cosmic impact beggars the imagination. The incoming body, typically several kilometers across, might weigh a hundred billion tons. Traveling at a speed of perhaps twenty or thirty kilometers per second, it delivers a blow equivalent to at least a hundred million megatons of TNT, far more than all the world's nuclear weapons put together. When it enters the atmosphere, the object displaces a vast column of air, creating a powerful shock wave that circles the globe. On hitting the ground, the meteor, along with much of the material at the impact site, is instantly vaporized. Huge quantities of rock are excavated from the surrounding terrain and hurled high into the air, even into space, leaving a gigantic crater. Large chunks of ejected rock rain back down again, hundreds or even thousands of kilometers away, glowing fiercely and igniting vegetation. The ground shock produced by the primary impact exceeds the most violent earthquakes, wreaking still more damage. If the meteor falls into the sea, tsunamis many kilometers high devastate the ocean rim, inundating immense tracts of land. The dust thrown up by the impact blankets the planet, blotting out the Sun for months, poisoning land and sea with acid rain. The deadly aftermath proves too much for many living species, and they are quickly driven to extinction.

The collision that created Lake Acraman was by no means an isolated event. Every few million years a comet or asteroid hits Earth with enough force to cause global devastation. In the past, such encounters would have been more frequent. It is becoming increasingly clear that cosmic impacts have had a major influence in shaping the evolution of life by triggering mass extinctions. It turns out, though, that cosmic impacts have not just altered the path of evolution; they also played a crucial role in the origin of life. Until recently, scientists appealed mainly to chemistry and geology in their attempts to explain biogenesis. Earth was treated as an isolated system. But over the last decade the crucial importance of the astronomical dimension of life has sunk in. To understand how life began, it seems we must look to the stars for answers.

The stardust in your eyes

> *If atom stocks are inexhaustible,*
> *Greater than power of living things to count,*
> *If Nature's same creative power were present too*
> *To throw the atoms into unions—exactly as united now,*
> *Why then confess you must*
> *That other worlds exist in other regions of the sky,*
> *And different tribes of men, kinds of wild beasts.[1]*

With these stirring words, the Roman poet-philosopher Lucretius seeks to persuade us that we are not alone in the universe. Lucretius reasoned that, if the universe were made of identical atoms subject to universal laws of nature, then the same processes that produced life on Earth should also produce life on other worlds. The argument, which dates back to the Greek atomist Epicurus, is compelling. But is it correct?

Astronomers have confirmed from spectroscopic observations that atoms are indeed the same throughout the cosmos. A carbon atom in the Andromeda Galaxy, for example, is identical to one here on Earth. Five chemical elements play a starring role in terrestrial biology: carbon, oxygen, hydrogen, nitrogen, and phosphorus. These elements seem to be among the most plentiful in the universe.

Carbon is the truly vital element. It qualifies for pride of place because of a unique chemical property: carbon atoms can link together to form extended chain molecules, or polymers, of limitless variety and complexity. Proteins and DNA are two examples of these long chain molecules. If it wasn't for carbon, life as we know it would be impossible. Probably any sort of life would be impossible.

When the universe began with a big bang, carbon was completely absent. The intense heat of the cosmic birth precluded any composite atomic nuclei. Instead, the cosmic material consisted of a soup of elementary particles such as protons and neutrons. A majority of the protons remained unattached, and went on to form the nuclei of hydrogen atoms. However, as the universe expanded and cooled over the first few minutes, nuclear reactions transmuted some of the hydrogen into helium, and just a smidgin into carbon.

Most of the carbon in the universe comes not from the big bang, however, but from stars. Stars are nuclear fusion reactors that normally burn hydrogen to produce helium. In large stars, the next step is to convert helium into carbon. After that, other familiar elements—oxygen, nitrogen, and so on—get made. Most of these heavier substances remain confined in the stars, but occasionally some are released when a star explodes. There is also a steady stream of material blown out by the Sun in the solar wind, and a similar process will occur in other star systems. Either way, the dispelled substances mingle with the clouds of mainly hydrogen gas that roam interstellar space. In the fullness of time, should these gas clouds contract to form new stars and planetary systems, the carbon and other elements from dead stars will be mixed in with them.

Imagine our own solar system forming this way, four and a half billion years ago. A massive cloud of hydrogen, laced with heavy elements, gradually shrinks. Here and there, gravity tugs the gas into dense spinning blobs. These agglomerations of matter are destined to become a cluster of new stars. One such star is our Sun. Around it, gas and dust swirl in complicated patterns, forming a disk-shaped nebula. The light material drifts to the periphery of the nebula and eventually condenses into the giant gas planets, such as Saturn. The heavier elements concentrate in the inner regions of the disk, where they become incorporated into Earth and its neighbors. The material that makes up our planet is therefore not primordial, but the nuclear ash from stars that blazed and died long before the solar system even existed.

Since the Earth formed, its material has not remained inert. Carbon, hydrogen, nitrogen, and oxygen are continually recycled through the atmosphere and crust by geological and biological processes. When an organism dies and decays, its atoms are released back into the environment. Some of them eventually become part of other organisms. Simple statistics reveal that your body contains about one atom of carbon from every milligram of dead organic material more than a thousand years old. This simple fact has some amazing implications. You are, for example, host to a billion or so atoms that once belonged to Jesus Christ, or Julius Caesar, or the Buddha, or the tree that the Buddha once sat beneath.[2]

Next time you look at your body, reflect on the long and eventful history of its atoms, and remember that the flesh you see, and the eyes you see them with, are literally made of stardust.

Cosmic chemistry

I grew up believing that chemistry is something that happens in test tubes. It therefore came as something of a surprise when I learned, in 1969, that molecules of ammonia and water had been discovered in outer space. How did they get there? I wondered. Of course, astronomers have known for a long time that space is not completely empty. The gaps between the stars contain tenuous clouds of gas and dust. But even a dense interstellar cloud can boast only a million atoms per cubic centimeter, which would be considered a hard vacuum in a laboratory. With so diffuse a medium, and sub-sub-zero temperatures, there would seem to be little scope for chemical reactions to take place. But not so.

Historically, the first hint that there may be molecules in space dates from the early 1920s, when an astronomer named H. L. Heger discovered some odd features, called "diffuse interstellar bands," in the spectra of stars. They were eventually put down to absorption by unknown molecules lying in space along the light path, but the idea didn't catch on. Decades later, following the unexpected discovery of interstellar ammonia and water, the list of known molecules in space began to grow rapidly. Today, over one hundred chemicals have been identified, mostly using radio and infrared telescopes.

Many of the interstellar molecules are organic. Commonest is carbon monoxide, but acetylene, formaldehyde, and alcohol are also plentiful. More complex organics, such as amino acids and PAHs (polycyclic aromatic hydrocarbons—of which, more later), have also been detected. It is now clear that not only are the basic life-encouraging elements abundant throughout the universe, so are many of the organic molecules actually used by life. With billions of years available for cosmic chemistry to generate these substances, there has been plenty of time for them to build up in the giant molecular clouds from which stars and planetary systems emerge.

Astronomers who study the chemistry of interstellar gas clouds are convinced that dust particles play an important role. Chemicals attach to their solid surfaces and react in complicated ways. It isn't hard to spot dust in space. Glance at the night sky near the constellation of Cygnus and you will notice great black blotches in the Milky Way. These dark areas are created by large clouds of dust that block out the starlight from beyond. The culprits are very tiny grains—typically a thousandth of a millimeter across, but extending down to molecular size. Their composition is the product of many physical and chemical influences—ultraviolet radiation, stellar winds, shock waves, cosmic radiation. They include silicates, ices, and carbonaceous material such as graphite, as well as many organics. Interstellar clouds can be many light-years across, so the total mass of dust in them is enormous. Tiny they may be, but interstellar grains could be the unwitting chemists that spawned life.

Curiously, interstellar dust has effects even in our own cosmic backyard. The inner solar system is a surprisingly dusty place, as space probes have discovered. The famous zodiacal light, visible after sunset in tropical latitudes, is caused by the scattering of sunshine by minute particles in space. Much of this material is home-grown debris, but some of it is streaming in from interstellar space. You can tell the particles that hail from the stars by their speed. Duncan Steel, formerly at The University of Adelaide, and his colleagues have used a ground-radar system in New Zealand to study interstellar grains that hit the Earth. By analyzing the trails of ionization created when micrometeorites plow into the atmosphere, these researchers worked out that some of them have speeds in excess of seventy kilometers per second—too fast for them to be trapped in orbit within the solar system.[3]

Genesis from space

The *Pioneer 10* spacecraft was launched from Cape Kennedy on March 2, 1972. It slipped out of radio contact on April 1, 1997, when it was ten billion kilometers from the Sun, making it the most remote man-made object then in existence. Imagine that you have hitched a ride on board *Pioneer 10* for a tour of the solar system and beyond. In six months, you cross the orbit of Mars and go on to negotiate the asteroid belt successfully. In late 1973, you pass close to Jupiter. Ten years later, you cross the orbit of Neptune and leave interplanetary space forever, heading for the stars. Already the Sun appears only one-thirtieth of the size it looks from Earth, and it is shrinking all the time. Ahead lies a chasm of emptiness, cold and dark. The nearest star is 4.3 light-years away—forty trillion kilometers. Even if you were going that way, which you aren't, it would take ten thousand years to get there at this speed. Settle down for quite a wait. There isn't going to be much to see for a long while.

After you have been traveling through space for several thousand years, and the Sun is reduced to an object no brighter than a very bright star, there is a flurry of activity close by. Something is out there in the blackness of interstellar space. A dark lump of matter suddenly looms up and glides past. Roughly spherical in shape, it measures ten kilometers across. On closer inspection the object is revealed to be an untidy jumble of rocks, ice, and tar: a comet.

As you journey on, more and more comets appear, and slip silently by. You are sweeping through a cloud of these elusive objects, a trillion dirty snowballs altogether, clustered in a swarm that envelops the Sun and planets. Here, fully a light-year from the center, this vast assemblage of minor bodies marks the true outer limit of the solar system. Far-flung they may be, but the comets remain bound by the Sun's feeble gravitational field.

Nobody has actually seen the cloud of comets surrounding our solar system, but its existence has been accepted by astronomers since Jan Oort first predicted it in 1950. The inert lumps of matter in the Oort cloud don't resemble the comets of popular lore, which shine in the sky and sprout tails. But the Oort cloud is the comets' true home, and it provides an almost inexhaustible reservoir of them.

Comets remain something of an enigma, even though they have been closely observed for centuries. Until recently, most astronomers dismissed them as spectacular but minor players in the celestial drama, in spite of the dread and foreboding that their passage engendered in past cultures. But opinion has begun to shift; comets are now a hot topic. One reason for this concerns their age. They are true relics from the birth of the solar system, near-pristine samples of solar-nebula material, perhaps infused with interstellar matter of even greater vintage. The dust emitted by Comet Halley, for example, is thought to be the most primitive substance ever analyzed by scientists. Deep-frozen in the depths of space, this primordial comet-stuff has been preserved largely unchanged for four and a half billion years.

Of more pressing interest is the role that comets seem to have played in the origin and evolution of life. To understand their importance, it is necessary to hark back to the beginning of the solar system. The way in which the planets formed from the turmoil of the solar nebula was rather complicated. The process started with the aggregation of tiny grains. These particles in turn collided and merged, slowly forming larger and larger lumps of solid material. In the inner solar system the grains were mainly heat-resistant silicates. Farther out, more volatile substances, including organics, condensed.

As the fragments grew in size and mass, so they began to exert a gravitational pull on their neighbors. Collisions became increasingly violent as the bodies were drawn together with greater force.

After perhaps ten thousand years of milling about, these objects would have swollen into planetesimals some hundreds of kilometers in size; after a million years, scores of Mars-sized planets were orbiting the Sun. Encounters of awesome magnitude became unavoidable. At some point, the incipient Earth was struck obliquely by one such object, with profound consequences. The giant body plowed to the center of our planet, and created its iron core. The lighter mantle was ripped away into space by the force of the collision, giving Earth its own mini-disk of orbiting debris that soon accumulated to form the Moon. The enormous energy of this cataclysm baked the Earth dry of any volatile material.

Farther out in the solar system, the pace of events was less frenetic, because the material there was more tenuous. The cooler conditions allowed substances like water and sulfur to solidify. Crucially, delicate hydrocarbons from the original gas cloud would have survived the proto-Sun's heat in this region. Tiny dust grains gathered fluffy snow as they swept up ice crystals. From time to time the snowflakes collided and stuck together. Being more widely scattered, these icy particles did not readily aggregate into planets, but instead formed a legion of minor icy bodies, ranging in size from comets a few kilometers across, to icy planetesimals a hundred times bigger. After about ten million years, enough of these icy bodies came together to create the embryo of the giant planet Jupiter. Once a critical size of about ten Earth masses was reached, Jupiter began to grow by runaway accretion. Its powerful gravitational field sucked in or scattered away debris from a wide band of the nebula, robbing the asteroid belt of enough material to form a separate planet, and condemning Mars to its dwarf status. The same pattern of growth was repeated for Saturn, Uranus, and Neptune, but at a slower rate because of the lower density of the nebula farther out. Beyond the orbit of Neptune, the planetesimals were too sparsely distributed to make any planets. (Pluto is not a true planet.) A lot of the icy planetesimals are still out there on the periphery of the solar system, dim and inconspicuous, orbiting the Sun in what is known as the Kuiper belt.

Over the eons, the gravitational fields of the giant outer planets have flung many small icy bodies deep into interstellar space. Most of them were ejected from the solar system altogether, never to return; others were hurled only as far as what eventually became the Oort cloud. This gravitational scattering was completely haphazard, and millions of icy lumps were propelled towards the inner solar system too, some of them to crash into the planets. Earth was repeatedly struck, first by asteroids from the region between Mars and Jupiter, then by comets from Jupiter's zone. Over a longer time scale, icy bodies from farther out, displaced by Saturn, Uranus, and Neptune, also smashed into the inner planets. These bodies from the outer solar system added a veneer of light rocky material to the Earth's crust. More significantly, they also delivered vast quantities of water, enough to make the present oceans many times over. And along with the water came many other volatile substances that the nascent Earth lacked, especially life-encouraging organics. By this stage the hydrogen, helium, and other gases of the original solar nebula had been blown away by a fierce solar wind, some to accumulate in Jupiter's atmosphere, most to be lost to interstellar space. Quite likely Earth was left with little or no primary atmosphere at all. But with the influx of cometary material, the planet became cloaked once more in a dense blanket of gases, augmented by volcanic vapors pouring from the molten interior.

After one hundred million years, the formation of Earth was more or less complete. However, for over half a billion years more, it would bear little resemblance to the serene blue planet we know today. The surface was hot, the oceans were much deeper, and the atmosphere was crushing. Volcanism was extensive, the Moon closer, and the tides were huge. The planet spun a lot faster then than now; day followed night in just a few hours. But the biggest difference concerned the continuing threat from space. The asteroids and comets that had helped fashion the surface conditions on the young planet didn't abruptly cease their activities. They kept on coming, age after age, with their cargoes of ice and organic material. In fact, they are still coming. Their contribution to the history of life had only just begun.

Impact

Comets gave it and comets taketh it away.

CARL SAGAN[4]

One of the reasons the Bible is such a good read is that it is full of drama and spectacle: fire and brimstone, signs in the heavens, floods, parting waters, plagues and pestilences. If the world was created six thousand years ago, as many Christians once believed (and a few apparently still do), God would have been busy indeed shaping the present form of our planet, building mountains and oceans, scouring valleys, moving glaciers.

When the geologists of the eighteenth century tried to explain mountains and river valleys, salty oceans and glaciation, rock strata and fossils in terms of physical processes rather than divine action, they realized that it would take far longer than six thousand years for these features to form. In 1785, James Hutton, the Scotsman epitaphically described as the founder of modern geology, declared of Earth's geological history, "We find no vestige of a beginning—no prospect of an end."[5] Hutton believed that the surface features of the Earth were shaped gradually by incremental changes extending over enormous lengths of time. He realized that millions of years would be needed to accumulate rock sediments and to raise and erode mountains.

Hutton's ideas became known as uniformitarianism, in contrast to the catastrophism of more biblically minded scholars, who sought explanation for Earth's form in Noah's flood, volcanic mayhem, and celestial thunderbolts. Hutton's thesis was wholeheartedly embraced by Charles Lyell, who carried the message of uniformitarianism to the people in a book, published in 1830, entitled *Principles of Geology*. By this stage it was becoming clear to scientists that geological changes would probably require billions, not merely millions, of years to complete. Such a conclusion well suited Charles Darwin, who envisaged biological evolution as a long series of slow adaptations, accumulating over similarly vast lengths of time.

With hindsight, we can see that uniformitarianism was ideologically driven, a reaction against religious interpretations of nature. As a result, it has proved a remarkably stubborn doctrine. Evidence for sudden geological and biological upheavals was obvious for a long time, yet it was largely ignored. Those who drew attention to it tended to be dismissed as cranks. When the respected astronomer Edmond Halley surmised in 1694 that a comet may occasionally strike a planet, his suggestion was shrugged aside. In 1873, the British astronomer H. A. Proctor was daring enough to propose that the lunar craters might be the result of impacts by meteorites, but he quickly withdrew the claim, citing the apparent absence of similar craters on Earth. Even in the 1960s, some astronomers were sure that lunar craters were mostly volcanic in origin. It took the Apollo landings to prove finally that the Moon's craters were actually produced by an extended bombardment from space.

Photographs of other planets and moons show a similar picture of heavy cratering: Mercury and Mars provide excellent examples. These bodies have preserved the record of collisions because they lack a thick atmosphere and have little geological activity. By contrast, most of Earth's impact craters have been obliterated by erosion. But not all. At least twenty-five impact sites have been positively identified in Australia alone. The United States has one of the most celebrated craters, near the town of Winslow, Arizona. Known as Meteor Crater or the Barringer Crater, it is 1.2 kilometers across, one hundred meters deep, and thirty thousand years old. Considerably older and bigger impact craters are known, such as the one at Lake Acraman I have already mentioned.

The best way to reconstruct the bombardment record of Earth is to study the Moon. Astronomically, it is so close that we can be sure Earth would have been a target for whatever treatment was meted out to our junior neighbor. And that was pretty severe. Even a pair of binoculars will reveal some of the larger lunar craters. The oldest date back over four billion years. Many smaller ones are more recent. Craters from later impacts are often found superimposed on earlier scars. Because big impacts tend to erase the earlier record, filling in the details of the first five hundred million years is largely a matter of guesswork. Mathematical models and the surviving lunar record suggest that all the bodies in the inner solar system were pummeled both by local debris and by asteroids and giant comets coming in from the outer solar system. This bombardment gradually tailed away over a few hundred million years, only to resume with renewed ferocity between about 4 billion and 3.8 billion years ago. It was this later phase of intense violence that created the familiar lunar maria—the dark, flat basins that were filled with lava and left relatively smooth in the quiescent aftermath. Opinions differ as to the cause and extent of the late heavy bombardment. Some astronomers believe that it was restricted to the vicinity of Earth, others that it encompassed the entire solar system. It may have been caused by the breakup of a moon or a monster comet.

From the point of view of life, the significance of this intense barrage was the delivery of organics. When the *Giotto* spacecraft flew close to Comet Halley in 1986, it revealed a tar-black core containing carbon, hydrogen, nitrogen, and sulfur. Analysis of the dust grains streaming from the head showed that as much as one-third was organic material. Common substances such as benzene, methanol, and acetic acid were detected, as well as some of the building blocks of nucleic acids. If Halley is anything to go by, comets could easily have supplied Earth with enough carbon to make the entire biosphere. A similar picture is emerging for the much larger icy bodies on the fringes of the solar system. Astronomers recently found several curious objects, dubbed Centaurs, which have wandered in from the Kuiper belt. These planetesimals are dark red, and seem to be covered with primal gunk rich in hydrocarbons.

It is tempting to conclude that impacting comets, asteroids, and planetesimals coated an initially barren Earth with a layer of organic matter and water, thus forming the primordial soup from which life eventually emerged. There is, however, a complication with this

156 o THE FIFTH MIRACLE

theory. The impact of a comet is a very violent event, more likely to destroy organic material than to deliver it. Small objects entering the atmosphere at high speed tend to burn up completely; large projectiles hit the ground with such force that they are mostly vaporized by the explosion. For organics to survive, the incoming object has to get lucky. As we shall see in chapter 9, with the right projectile mass and angle of entry, extraterrestrial organic matter can reach the ground unscathed, but it is unusual. Some researchers believe that dust grains are likely to fare better than large rocks, and that most of Earth's organics floated down from the sky in this particulate form, like manna from heaven. Others think that shock waves from incoming comets will generate organic molecules afresh to offset what the impacts destroy.

In the extreme case, a very large body will hit the Earth with such violence that it removes more material than it deposits. This is euphemistically referred to as impact erosion. It seems that the larger collisions during the heavy-bombardment era had enough punch to strip away much of the atmosphere and oceans. Cometary bombardment is therefore a two-edged sword when it comes to water and organics. Whether a planet is a net winner or loser depends very much on the circumstances. It appears that small bodies like Mars, Mercury, and the Moon lost out to impact erosion, whereas Earth and Venus have on balance gained material.

The two-edged sword remains, poised like the sword of Damocles over our planet. Comets, asteroids, and meteors continue to menace Earth. The reason for this can be traced to influences way beyond the solar system. Although to human eyes the stars seem to be fixed in the sky, they and our own Sun are, in fact, in orbit around the galaxy, completing one circuit every 250 million years or so. As a result of this slow migration, from time to time another star or a massive cloud of gas approaches the solar system. When this happens, the gravitational field disturbs the Oort cloud. Some comets are tossed out of the solar system, others are deflected back towards the planets.

When a new comet is discovered, the chances are that it is a one-time visitor, dropping in on us from the outer reaches of the

Oort cloud on a multimillion-year journey. Sometimes when a comet sweeps through the inner solar system, its orbit is perturbed by Jupiter or another planet, so that it returns periodically. Many periodic comets are known, the most famous being Halley's. When they approach the Sun, the volatile material starts to evaporate, and the comet emits a cloud of gas and dust that is drawn by the solar wind into the distinctive tail. The ultimate fate of these objects is either to fall into the Sun, to hit a planet, or to get ejected once more from the inner solar system. Alternatively, the comet may "die"—i.e., lose all its volatile material, cease to glow, and disintegrate—before any of these events occurs.

Calculations indicate that disturbances to the Oort cloud should displace most of its comets after a few hundred million years. As comets still appear regularly, some replenishment process must be at work. Astronomers suspect that there is an inner cloud, or feeder belt, responsible for this, extending from the region beyond Neptune and tapering off gradually, containing a total of about two Earth masses. Just in the last few years, several large icy bodies have been discovered near or beyond the planetary edge of the solar system, in the Kuiper belt. Probably many short-period comets originate here rather than in the more distant Oort cloud.

Even today a comet or asteroid could hit Earth with enough force to destroy most life. It now seems likely that massive collisions have caused several major annihilation events over geological time. The most famous mass extinction occurred sixty-five million years ago (relatively recently in geological terms), when the dinosaurs suddenly died out, along with a large number of other species. Evidence that a huge cosmic impact was responsible comes from the discovery of a worldwide layer of the rare element iridium, deposited in clay strata laid down at that time. This iridium was almost certainly delivered by the impactor. Dramatic confirmation of the theory came in 1990, with the discovery of a gigantic crater of the right age buried under limestone in Mexico. It measures at least 180 kilometers across, and was probably made by an object about 20 kilometers in diameter.

Cosmic impacts are examples of what biologists refer to as contingent events. They take no account of terrestrial biology. They just happen, out of the blue, without any causal connection to the evolution of life on Earth. They are both creative and destructive, good and bad. The origin of life on Earth—and perhaps other planets too—may well have depended on their volatile-rich material; the death of the dinosaurs served to clear the way for the ascent of mammals and, eventually, mankind. It seems we owe our very existence to a chance astronomical catastrophe. Whether mankind will someday go the way of the dinosaurs remains to be seen.

The Sisyphus effect

The discovery that Earth and Moon endured a punishing cosmic barrage until 3.8 billion years ago presents us with a major puzzle. If the fossil record is to be believed, life was certainly flourishing 3.5 billion years ago, and quite possibly as early as 3.85 billion years ago. Given the dire consequences of a major impact, could life have endured through the late heavy bombardment? Unfortunately, the trail of evidence goes cold just as this problem gets interesting. Although geologists have found isolated zircon crystals 4.2 billion years old, and inferred that some sort of solid crust must have existed at that time, the oldest intact rocks ever found date back 4.03 billion years. Geological processes have eradicated almost all evidence of what our planet may have been like before about 3.8 billion years ago.

Though Earth may be reluctant to yield the secrets of her youth, indirect evidence for conditions prior to 3.8 billion years ago may lie right under (even within!) our noses. The DNA of our bodies contains a record of the past, because our genes have been fashioned by environmental circumstances. Though the genetic record, like the geological record, has been garbled and obscured by the ravages of time, it is not completely erased. By prizing out information from genes, microbiologists can tell a lot about the universal ancestor that may have lived some four billion years ago, and with this information we may guess something about the conditions that prevailed at the time. The message that emerges is quite a surprise.

Imagine what it was like during the epoch of heavy cosmic bombardment. Every large impact created global upheaval. The scale of the mayhem was far worse even than the dinosaur-destroying blast. As late as 3.8 billion years ago, the Moon was hit by an object ninety kilometers in diameter, producing a colossal impact basin the size of the British Isles. Several similar cataclysms have left signs of their damage in the form of mountain rings. Being that much bigger, Earth must have suffered dozens of collisions of this magnitude, as well as some that were even larger. Nor are the culprits for these mega-impacts hard to find. There are many large bodies lurking in the solar system even today. Chiron, a recently discovered planetesimal, is on an unstable orbit near Saturn and measures 180 kilometers across. The consequences of its hitting the Earth are too horrible to contemplate. And Chiron is by no means the largest known minor planet. Four billion years ago, such objects would have been far more common than they are today.

The dramatic effects of massive collisions have been analyzed by Norman Sleep and his colleagues at Stanford University.[6] An impactor 500 kilometers in diameter would excavate a hole 1,500 kilometers across and at least 50 kilometers deep. A huge volume of rock would be vaporized in a gigantic fireball that would spread rapidly around the planet, displacing the atmosphere and creating a global furnace. The surface temperature would soar to more than three thousand degrees Celsius, causing all the world's oceans to boil dry, and melting rock to a depth of almost a kilometer. As the crushingly dense atmosphere of rock vapor and superheated steam slowly cooled over a period of a few months, it would start to rain molten-rock droplets. A full millennium would elapse before normal rain could begin, presaging a two-thousand-year downpour that would eventually replenish the oceans and return the planet to some sort of normality.

Although there may have been only a few catastrophes as severe as this, Sleep estimates that there must have been hundreds comparable with the events that made major lunar features such as Mare Continentale. These would splash molten rock into space and create a transient rock-vapor blanket above Earth's atmosphere. The radiant heat beating down from the sky would be enough to boil away the top forty meters of ocean and trigger decades of scalding rain.

Clearly, large impacts have the effect of thoroughly sterilizing the Earth's surface. The searing heat pulse from the rock vapor would destroy any exposed organisms in pretty short order. If Earth was pounded as fiercely as astronomers believe, and if surface organisms really were well established by 3.8 billion years ago, then life must have burgeoned almost as soon as the effects of the last sterilizing impact were over. This suggests either that life came from space, or that it emerged quickly once conditions were halfway reasonable. (Of course, with a sample of one, it is hard to be too confident in this conclusion.) Either way, it makes sense to consider the possibility that life might have got going more than once. The late heavy bombardment may have been preceded by a relatively quiescent phase. In any case, the barrage must have tailed off gradually as the projectiles became depleted, leaving gaps of varying duration between successive sterilizing impacts. These gaps would have provided windows of opportunity for life to arise.

A few years ago, Kevin Maher and David Stevenson of Caltech sought to redefine what is meant by the origin of life in the light of the bombardment scenario.[7] Life could be said to have started, they reasoned, when the time it took for self-replicating organisms to emerge was less than the time between sterilizing impacts. If it took, say, ten million years to make life from a primordial soup, the bombardment would have needed to leave at least ten-million-year windows in order for life to begin. Maher and Stevenson then asked how far back you could go into the bombardment era and still expect gaps of that duration. They came up with an answer of two hundred million years. So life might have arisen at any time after about four billion years ago, flourishing in the calmer periods, only to be wiped out again by the next sterilizing impact. Like the mythical Sisyphus, condemned to keep rolling the stone up the hill only to fall back again each time, life may have struggled over and over to establish itself, only to get zapped repeatedly from space.

It is a curious thought that, if life did form anew several times, then humans would not be descendants of the first living thing. Rather, we would be the products of the first life forms that just managed to survive the last big impact in this extended stop-go series. Which raises an interesting point about the 3.85-billion-year-old rocks at Isua. A sterilizing impact could have occurred *after* life had transformed them. If so, the organisms that left their subtle traces in that ancient terrain may not be ancestral to our form of life at all. They may have belonged to an earlier, alternative biology that was totally wiped out by the cosmic bombardment. The rocks of Greenland may thus contain evidence for what is, in a sense, an alien life form.

From what we know of the early history of the solar system, the Earth's surface was a hazardous place for a living organism to be for at least several hundred million years after the planet's formation. Even the bottom of the ocean would afford little protection against the violence of the larger impactors. The heat pulses from these cataclysms would have been lethal to a depth of tens or even hundreds of meters into the Earth's crust itself. Hardly a Garden of Eden. Where, then, would one expect the earliest life forms to have taken up residence? What refuge existed that might have spared the first faltering ecosystem wholesale annihilation by vaporized rock? The answer would seem to be: somewhere deep. Somewhere below ground.

But what on Earth can live there?

CHAPTER 7

Superbugs

IN THE LATE 1920S, the Egyptian capital of Cairo was plagued by a spate of main drain collapses. Investigations revealed that the concrete lining of the sewer pipes had simply disintegrated after as little as two years in the ground. Civil engineers began a series of experiments to determine the cause of the damage. Rapidly crumbling sewers then began to crop up in other places too. In Orange County, California, the twenty-six-mile trunk outfall became badly corroded, and had to be chlorinated to stop the rot; the fifty-five-mile northern outfall in Los Angeles was prevented from total collapse only with the aid of forced ventilation. In South Africa, Cape Town engineers were baffled by galloping corrosion of their concrete sewer pipes, some of which were devoured at the rate of a quarter-inch per year. Clearly something strange was going on below ground.

When the sewers in several Australian towns and cities also began collapsing, the Melbourne and Metropolitan Board of Works were called in. A research project was set up under special investigator Dr. C. D. Parker, who obtained samples of severely affected sewer pipes from around the country. By that stage, engineers already suspected that the problem was somehow associated with hydrogen sulfide—the evil-smelling gas reminiscent of bad eggs—but the sheer speed and virulence of the corrosion was puzzling.

It was not long before Parker discovered what was happening.[1] Previous theories focused on some sort of chemical transformation of the concrete, but Parker realized that the corrosion was in fact due to biological attack. He soon isolated the culprit: a slender rod-shaped bacterium about two micrometers long. This bizarre microorganism eats into solid concrete, turning it into a puttylike substance after just a few weeks. Unlike normal organisms that consume organic matter to grow, Parker's microbes seemed to thrive on a diet of sulfur, which they extracted from the hydrogen-sulfide gas emitted by the sewage. Parker was able to culture the bacteria, and he chose the tentative name *Thiobacillus concretivorus*, meaning "concrete-eating sulfur rod."

Laboratory tests revealed that *Thiobacillus concretivorus* produced sulfuric acid, and it was this that was destroying the concrete sewer pipes. Indeed, the isolated bacteria refused to grow unless they were immersed in sulfuric acid. The concentration of the acid was astonishing, enough to kill all other creatures and even strong enough to dissolve strips of metal! It turned out that Parker's acid-loving bugs were already known to science; they had been discovered many years earlier and given the name *Thiobacillus thio-oxidans*. They are one of a number of micro-organisms known as acidophiles—acid lovers—that positively demand an acid medium in which to live, and lurk in places like coal and iron-ore dumps. Some of them can tolerate a fluid with a pH as low as 2, which would prove distinctly painful if you were to put your hand in it.

No less remarkable than *Thiobacillus thio-oxidans* is a tough little microbe called *Halobacterium halobium*, found where no life is supposed to exist—in the Dead Sea. This inland lake is so salty that bathers can easily sit upright (I once tried it myself). The high salt content is due to the sea's being landlocked. Water flows in from the River Jordan and then evaporates, leaving the salt. The area around the Dead Sea is dry and barren, much of it resembling a moonscape.

In spite of its forbidding setting and suggestive name, the Dead Sea is not completely dead, as the discovery of *Halobacterium halobium* attests. Nor is it unique in providing a habitat for salty bugs, which are known collectively as halophiles. The Great Salt Lake in Utah and Lake Magadi in Kenya play host to their own microbial inhabitants. Viable halophiles have also been found in salt mines and entombed in ancient crystals.

Microbes are known that can survive other extremes too, such as intense cold. Bacteria have been found thriving in water trapped beneath the Antarctic ice sheet, for example. Some can tolerate being cooled to liquid-nitrogen temperatures, or even lower. Other micro-organisms live happily in extremes of alkalinity. *Plectonema*, for example, will grow in a solution so alkaline that it would seriously damage human skin. There are even bacteria, like *Micrococcus radiophilus*, that put up with radiation that would quickly prove lethal to most other organisms. Indeed, healthy microbes have been found inhabiting the waste tanks of nuclear reactors, ingesting uranium, plutonium, and other radioactive elements. Nor is pressure an obstacle. Common bacteria such as *E. coli* can be subjected to several hundred atmospheres without undue harm. At the other extreme, viable specimens of the bacterium *Streptococcus mitis* were retrieved from the surface of the Moon, where they endured a complete vacuum for two years while attached to a camera housing on the *Surveyor III* spacecraft.

The evocative terms "superbug" and "extremophile" have been coined to designate these hardy micro-organisms. At first superbugs were merely a scientific curiosity, studied mainly for their possible commercial exploitation. Recently, however, as microbiologists have extended their knowledge about them, these organisms have taken on an altogether more profound significance. Some superbugs seem to be extraordinarily ancient and primitive, and there is a growing feeling among scientists that they could be living fossils, the nearest thing alive to the universal ancestor. If so, the rigorous conditions in which they thrive, although extreme to us, might be indicative of what Earth was like 3.9 billion years ago.

Some like it hot

> *Organic life beneath the shoreless waves*
> *Was born and nurs'd in Ocean's pearly caves.*
>
> ERASMUS DARWIN[2]

In late summer the temperature in Adelaide, where I live, can sometimes hit 43 degrees Celsius, or 109 degrees Fahrenheit. When it does, most people stay indoors. Outside, keeping cool is a major problem. Even our cat has been known to pant like a dog. Some desert animals can tolerate temperatures a bit higher than this, but 50 degrees Celsius, or 122 degrees Fahrenheit, seems to be about the limit. Much hotter, and both animals and plants literally start to cook. Cooking has the effect of unraveling proteins so they can no longer function properly. The classic example is an egg, which turns white and solid when immersed in moderately hot water. If that sort of thing started to happen to a live animal, it would soon be dead.

Several decades ago, biologists were surprised to discover that certain bacteria live comfortably at temperatures up to 70 degrees Celsius—158 degrees Fahrenheit. These peculiar microbes were found in compost heaps, silage towers, and even domestic hot-water systems. For obvious reasons, they were christened thermophiles. Investigation revealed that thermophiles use special stabilizing proteins and are encased in cell membranes made of a type of heat-resistant wax rather than normal fat. For a time it was assumed that 70 degrees Celsius marked a strict upper limit to the temperature of thermophiles' habitats, beyond which even their DNA would start to melt. It therefore came as an even bigger surprise when, in 1969, Thomas Brock of Indiana University found a superbug, which he named *Thermus aquaticus*, living in the hot springs of the Yellowstone National Park at temperatures of 80 degrees Celsius, or 176 degrees Fahrenheit.

As it turned out, this was just the beginning. In the late 1970s, the submersible vessel *Alvin*, belonging to the Woods Hole Oceanographic Institute, was used to explore the seabed along the Galápa-gos Rift in the Pacific Ocean. This feature, some two and a half kilo-meters below the surface, is of interest to geologists as a prime example of submarine volcanic vents known as black smokers. The name derives from the appearance of the mineral-coated rock chim-neys from which dusky fluids billow forth into the surrounding ocean. Near a black smoker the seawater can reach temperatures as high as 350 degrees Celsius—way above the normal boiling point. This is possible because of the immense pressure at that depth.

To the astonishment of the scientists involved in the *Alvin* proj-ect, the region around the Galápagos black smokers, and several other deep-sea locations, turned out to be teeming with life. Among the more exotic denizens of the deep were crabs and giant tube worms. There were also familiar thermophilic bacteria on the pe-riphery of the black smokers. Most remarkable of all, however, were some hitherto unknown microbes living very close to the searing ef-fluent in temperatures as high as 110 degrees Celsius. Few scientists had ever seriously imagined that any form of life could withstand such extreme heat.

Organisms that grow fastest above 80 degrees Celsius were dubbed hyperthermophiles by Karl Stetter, who isolated and de-scribed many of the early examples. Following their discovery, it soon became clear that these superbugs are not freaks. To date, about twenty genera have been described. Significantly, many hy-perthermophiles are archaea (see chapter 3). The official tempera-ture record is currently held by an organism known as *Pyrolobus fumarii*, discovered by Stetter and his colleagues, that reportedly grows at 113 degrees Celsius. However, John Parkes of Bristol Uni-versity claims, somewhat amazingly, to have evidence of microbes living at temperatures as high as 169 degrees Celsius.[3]

A basic question about these deep-sea organisms is: how do they make a living? Biologists long supposed that all life on Earth de-pends ultimately on the Sun for energy. Plants won't grow without light, and animals must eat plants (or each other) to survive. How-ever, that far beneath the sea it is pitch-black.[4] No sunlight penetrates. This isn't a problem for the crabs and worms, because they scavenge for food among the smaller creatures on the seabed. But something must lie at the base of the food chain. It turns out that microbes act as primary producers, obtaining their vital energy directly from the hot chemical broth vomiting from the volcanic depths.

Organisms that don't eat organic matter but manufacture their own biomass directly are known as autotrophs ("self-feeders"). Plants are the most familiar autotrophs; they use the energy of sunlight to turn inorganic substances like carbon dioxide and water into organic material. Autotrophs that make biomass using chemical energy rather than light energy have been dubbed chemoautotrophs, or chemotrophs for short. The discovery of true chemotrophs was a pivotal event in the history of biology. Here was the basis of a completely independent life chain, a hierarchy of organisms that could exist alongside familiar surface life, yet without being dependent on sunlight for its primary energy source.[5] For the first time it became possible to conceive of ecosystems free of the complexities of photosynthesis. Scientists began to glimpse a vast new biological realm that has lain hidden for billions of years.

Life in the underworld

> *Microbial life exists in all the locations where microbes can survive.*
>
> THOMAS GOLD[6]

In his book *Journey to the Center of the Earth*, the famous science-fiction writer Jules Verne tells the story of an expedition into the Earth's interior. The intrepid explorers discover a whole new world beneath the ground, occupied by exotic life forms inhabiting subterranean caverns. Unfortunately, Verne's story flew in the face of the geological evidence of the day. Miners are well aware that deep means hot: the temperature can rise as much as twenty degrees Celsius for each extra kilometer you go down. This makes life intolerable for most organisms below a depth of a few kilometers. The temperature gradient continues into the Earth's crust, through its molten mantle, and into the core, at which the temperature rises to more than three thousand degrees Celsius. Any journey to the center of the Earth would mean certain incineration. Verne's dream that life might exist beneath the Earth's surface seemed ridiculous. Biologists have long been aware that topsoil contains bacteria and that limestone caves can be inhabited by specially adapted organisms. But apart from these exceptions, the planet was pronounced dead from the ground down.[7] However, the same opinion prevailed concerning the ocean depths. Nothing much could survive, it was thought, below the "photic zone"—the surface layers of ocean illuminated by sunlight. The discovery of black-smoker ecosystems changed all that. But if superbugs can exist several kilometers under the sea, might they not also exist several kilometers under the ground?

The first scientist to air publicly the view that life might be thriving deep below the Earth's surface seems to have been a Chicago geologist named Edson Bastin. In the 1920s, Bastin wondered why water extracted from oil fields contained hydrogen sulfide. He suggested that the gas might have been produced by sulfate-reducing bacteria living deep in the oil reservoirs. However, with little evidence to back his claim, Bastin found few supporters.

In fact, pointers to biological activity at great depth lay all around, if only geologists knew what to look for. In the 1960s, subterranean mineral deposits were discovered that appeared to have been precipitated by microbes. Iron, sulfur, manganese, zinc, and other substances known to be used by bacteria were concentrated in a suspicious manner. Meanwhile, Lloyd Hamilton, an Australian graduate student at the University of London, discovered unmistakable shapes of fossil microbes in veins of the mineral jasper. He concluded that these were remnants of iron-precipitating microbes that had made a home in the pores of rocks.[8]

In spite of the accumulating evidence for subterranean life, the prevailing opinion that the Earth's crust is sterile did not really begin to change until the late 1970s. At that time, governments were funding research into the problem of nuclear-waste disposal. Radioactive material had been buried in deep strata on the assumption that nothing much could happen to it. However, studies of groundwater had already hinted that bacteria might inhabit underground reservoirs, and rock samples returned from drilling surveys betrayed telltale signs of bacterial processing. It slowly dawned on scientists that if microbes could invade deep aquifers they might also get into underground nuclear dumps and corrode the containment vessels, eventually releasing the waste. Similar worries began surfacing in the petroleum industry, as it became clear that bacteria can also infiltrate oil reservoirs and sour the oil. But even by the late 1980s, most scientists were still resistant to the idea that life could flourish far beneath the Earth's surface. When Cornell University astrophysicist Tommy Gold announced he had found evidence for biological activity in Swedish granite nearly seven kilometers down,[9] his claim was initially greeted with derision.

It took the recovery of live micro-organisms to convince the skeptics. The U.S. Department of Energy commissioned an experimental drilling project in the Savannah River area of South Carolina, and the researchers began extracting rocks from a depth of half a kilometer with viable bacteria for everyone to see.[10] The project engineers took scrupulous care to avoid contaminating the samples with surface organisms, so there could be little doubt that the microbes really did inhabit the depths. Meanwhile, Karl Stetter's group managed to cultivate hyperthermophiles extracted from oil wells as deep as four kilometers.

These remarkable findings were confirmed by other drilling projects around the world. Three-kilometer boreholes drilled through Triassic sediment in the Taylorsville Basin in Virginia uncovered unique rod-shaped hyperthermophiles, including the imaginatively christened *Bacillus infernus*. The microbes at shallower locations tended to be mesophiles—organisms that grow in hot but not roasting conditions. Below two kilometers, thermophiles prevailed. Project scientists estimate that the Taylorsville site has been occupied by microbes for at least 140 million years. Some locations, like the hardrock Stripa mine in Sweden, are dominated by a handful of species; friable coastal sediments in South Carolina harbor communities containing hundreds of different varieties. The total inventory of deepliving microbe species currently runs into thousands. Some core samples have been obtained with up to ten million bacteria per gram. It is beginning to look as if the rocks beneath our feet are swarming with tiny life forms.

Now that the existence of subterranean superbugs has sunk in, scientists are rushing to rewrite the textbooks. All sorts of geological oddities are being attributed to the activities of unusual microbes. Acid-secreting bacteria, for example, can etch solid rock such as quartz, causing pitting and erosion. Perhaps this process is going on deep underground too? The networks of pores that enable oil to be extracted from sedimentary rocks might even owe their origin to these busy little organisms. If so, it opens up the lucrative prospect of harnessing superbugs to accelerate oil extraction.[11]

Groundwater movement is another target for bacteria hunters. Francis Chapelle of the U.S. Geological Survey in Columbia, North Carolina, has studied microbes at work in deep aquifers, and found that iron-dissolving bacteria can create pores and increase water flow, whereas sulfide-producing bacteria precipitate the dissolved iron again and close up the pores. He likens the microbes to minuscule lock-keepers, switching the flow on and off according to their requirements.[12]

A similar picture has begun to emerge in marine surveys. Not only do microbes live on and near the seabed, they also inhabit the sedimentary rock strata beneath the ocean floor. The international Ocean Drilling Program has recovered rocks showing signs of life from nearly a kilometer into the seabed. Samples from ten sites in the Mediterranean Sea and the Atlantic and Pacific Oceans have been studied by John Parkes and his colleagues in Bristol.[13] Again, meticulous precautions were taken to avoid the threat of contamination. Core samples were placed in a special sterile rig flushed with nitrogen, and their midsections excised using a hacksaw. The cut ends of the sample were then flamed and capped. Pretty well everything in sight was sterilized. The cores were stored in an oxygen-free environment at four degrees Centigrade (thirty-nine degrees Fahrenheit) until they could be analyzed in the lab a few weeks later. There the samples were further chopped up and peered at.

The results were sensational. The Bristol researchers found microbes in all the samples they studied, to a depth of 750 meters. The microbial colonies in the seabed were if anything more prolific than those found beneath the continents. Parkes was able to count the entombed bacteria under the microscope directly to confirm their astonishing fecundity. Populations ranged from more than a billion per cubic centimeter near the surface, to ten million deep down. Curiously, there is some evidence that the numbers start to rise again below a certain depth, with no end yet in sight. It is significant that about 5 percent of the bacteria retrieved were caught in the act of dividing, proving that they were alive and kicking when plucked from the deep. Indeed, some were still viable. Using a modified pressure-cooker, Parkes has been able to culture them in the laboratory.

It is clear from these recent discoveries that Earth possesses a pervasive living underworld, the vast extent of which is only just being revealed. There must be a huge amount of biomass in total down there. If bacteria proliferate to a depth of half a kilometer or more, as the surveys suggest, then, totted up over the whole planet, they would account for a tenth of the Earth's biomass. Even this could be an underestimate, because some types of microbe live happily at yet greater depths. If 110 degrees Celsius is as hot as they can stand, the microbial realm might go as deep as four kilometers under the ground and seven kilometers beneath the ocean floor. And if Parkes is to be believed, the top temperature might be as high as 170 degrees Celsius, and the habitable zone would go even deeper.

An obvious question to ask is how living organisms got to be in such deep locations in the first place. Did they infiltrate the rocks from above, swept along in the groundwater? Or did they get trapped long ago, when the sediments were first formed? It seems likely that both routes have been followed to some extent. However, these explanations proceed from the assumption that surface life is "normal," and subterranean life is an offbeat adaptation. Can we be sure of this? Could it be that the reasoning is literally upside down, and that the truth is just the opposite?

Ascent from Hades

Ever since Darwin's casual speculation that life started in some warm little pond, the conventional wisdom has been that life is and always was a surface phenomenon. The discovery of the hot, deep biosphere has dramatically altered that view. If life can flourish far beneath the Earth's surface, perhaps we should look downwards for the crucible in which the first living thing was forged.

There are several reasons why a location on the sea floor—or, better still, in the rock sediments beneath it—seems the most promising natural setting for the origin and early evolution of life. The most obvious concerns the cosmic-impact threat that I discussed in the last chapter. The violence of the late heavy bombardment would have effectively sterilized the Earth's surface repeatedly. With vaporized rock boiling the oceans and melting the land, conditions would have been lethal to a depth of tens of meters at least. But deeper down, micro-organisms could have withstood even the very large impacts. An added hazard of residing on the Earth's surface in the far past was ultraviolet radiation. Without an ozone shield, the sunlight would have proved deadly. Volcanic eruptions, more extensive then than now, would have belched forth enormous quantities of dust. Climatic variations due to aerosols and changes in atmospheric pressure brought about by the bombardment were probably extreme. Below the surface, however, conditions would have been far more stable and equable.

Another advantage of a deep location is that the raw materials needed for life were readily available. Even today, the Earth's crust exudes a steady supply of hydrogen, methane, ammonia, hydrogen sulfide, and other reducing gases. These are just the sorts of chemicals needed to synthesize biomolecules efficiently. In their famous experiment, Miller and Urey assumed that the Earth's primeval atmosphere was made up of such reducing gases, but now that geologists favor a mixture of carbon dioxide and nitrogen, the surface-soup theory doesn't look too encouraging. By contrast, in the subsurface realm, especially in the vicinity of volcanic vents, the crust would have provided a veritable abundance of reducing chemicals, including ferrous iron. Other nurturing substances, like sulfur and manganese, are also abundant in rocks and volcanic effluent. The spongy nature of sea-floor basalt helps by providing a labyrinth of channels and cavities to concentrate organic material and a vast surface area to catalyze reactions. In all, this adds up to a highly productive biochemical environment, an expectation borne out by experiment. Simulations of geothermally heated ocean crust yield far more organics than traditional Miller-Urey experiments.

Energy is another factor to consider, just as important as raw materials. Everett Shock of Washington University at St. Louis has computed the energy and entropy budgets near deep-sea hydrothermal vents. "There is an enormous thermodynamic drive to form organic compounds, as seawater and hydrothermal fluid, which are far from equilibrium, mix and move toward a more stable state," he explains.[14] Shock finds that the available energy is maximized at around 100–150 degrees Celsius, precisely the temperature range in which hyperthermophiles live. Not only can these organisms readily tap into the vast reserves of chemical and thermal energy provided, they can even *gain* energy by fabricating simple organic compounds. The energy released may then be used to pay for thermodynamically unfavorable reactions like peptide synthesis. Shock estimates that in a typical vent life can exploit this thermodynamic bonanza by creating biomass at the prodigious rate of two and a half kilograms per hour. This contrasts with the uphill struggle of photosynthesis, used by surface life, which demands special mechanisms to overcome the thermodynamic disadvantage. Whereas it is often said that there is no such thing as a free lunch, hydrothermal microbes get paid to eat lunch! "Nowhere else on Earth is the connection between geochemical and biological processes as profoundly evident as in hydrothermal systems," concludes Shock.[15]

Persuasive though these arguments seem, the most compelling evidence that life began hot and deep comes not from chemistry at all but from genetics. As I mentioned in the preceding chapter, the genes of extant organisms enfold a record of the past, and it is to molecular biology that we must look to discern the nature of the universal ancestor. What was it like and where did it live? The technique of gene sequencing pioneered by Carl Woese, which I introduced in chapter 3, can be used to reconstruct the tree of life and to determine the evolutionary distances between different microbes. From these studies it is possible to infer which group of organisms has evolved least over time, and is therefore most like early life. The results of this research point strongly to the archaea. The archaea, it will be recalled, constitute one of the three great domains of life. This domain split away from the others, the bacteria and the eucarya, a very long time ago, possibly as early as 3.8 billion years. But whereas most bacteria and eucarya have undergone substantial genetic changes, the evolutionary clock has ticked very slowly for the archaea.[16]

Among the many known archaean species, some groups stand out as the most sluggish of all at accumulating genetic change. These evolutionary stick-in-the-muds rejoice in names like *Pyrodictium* and *Thermoproteus*. Karl Stetter and Susan Barns have extensively studied such archaea using a technique called 16SrRNA analysis, which refers to a subunit of ribosomal RNA that can be extracted from uncultured organisms in the wild. Figure 7.1, based on earlier work by Woese and his colleagues, summarizes the latest results as a portion of the tree of life.[17] The most significant feature, the thing that leaps off the page, is that the lowest and shortest branches of the tree are dominated by thermophiles and hyperthermophiles. It is the organisms that cluster around the thermal ocean vents and inhabit the hot subsurface rocks that are the evolutionary throwbacks. The unmistakable message of the genes is that heat-

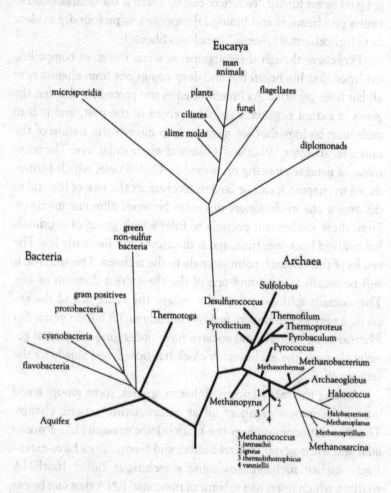

Figure 7.1. Living fossils. This portion of the tree of life depicts how far various species have separated genetically from each other. The lengths of the branches are proportional to the amount of genetic drift. The bold lines indicate hyperthermophiles discovered by Stetter and his group. Clearly the least evolved species, occupying the shortest and deepest branches of the tree, are all hyperthermophiles. (Reproduced from Gregory Bock and Jamie Goode, eds., *Evolution of Hydrothermal Ecosystems on Earth [and Mars?]* [Chichester: Wiley, 1996], by permission of the Novartis Foundation.)

loving, deep-loving microbes most closely resemble the universal ancestral organism.

Perhaps this is no surprise. Whereas the surface of the Earth has been transformed enormously over geological time, the subsurface has changed far less. There exist locations, such as sedimentary rocks on the sea floor and submarine hydrothermal vents, which are scarcely different from their counterparts billions of years ago. If life did begin in a hot, deep location, it may well have continuously occupied such environments up to the present day. With stable conditions, evolution would have stalled, and we might expect the inhabitants of these subsurface hot spots to differ little from their ancient forebears. The microbes that dwell beneath the ground and the ocean floor, and congregate in the scalding waters around hydrothermal vents, could therefore be hangovers from the turbulent epoch when life was striving to establish itself on a hot and dangerous planet.

When hyperthermophiles were first discovered, most microbiologists were inclined to dismiss them as aberrations—weird organisms that must have somehow invaded peculiar high-temperature niches, and evolved to cope with the unusual conditions. Now the evidence points to the opposite conclusion: the earliest microorganisms were all hyperthermophiles, and only later did some adapt to life at lower temperatures. In certain locations beneath the Earth's surface, pockets remain where conditions resemble those of very long ago. There one still finds organisms retaining a primeval lifestyle four billion years on. A black smoker may be a forbidding spot for you or me, but for an infernal organism like *Pyrodyctium occultum* it is a veritable paradise. Cosseted and content in their geothermal cocoons, these superbugs are actually biological wimps, clinging to the cradle of life, while all around them their more adventurous cousins have "gone forth and multiplied," having learned to cope with the harsher realities of life on or near the Earth's surface. If this theory is correct, the direction of microbial migration was up rather than down. Subterranean life didn't get buried, it was there from the start. Life ascended from the depths.

The theory that life began hot and deep was first mooted in

1981 by Jack Corliss of the University of Maryland,[18] and popularized by Tommy Gold in a trail-blazing paper published in 1992.[19] Initially greeted with considerable skepticism, this theory is rapidly gaining in popularity among scientists of many disciplines. Several lines of evidence from molecular biology support it. For example, hyperthermophiles assimilate carbon in a strange way, using a simple and rather primitive chemical cycle. The deepest-rooted organisms in the tree of life all use special heat-shock proteins that protect against sudden temperature fluctuations of the sort expected near volcanic systems. These proteins contain metals like zinc and molybdenum that are common in volcanic effluent. Support also comes from a detailed analysis of the temperature habits of microbes. The archaea, as we have seen, include many hyperthermophiles. The eucarya, which are generally more evolved and complex than the archaea, can muster only a handful. As for bacteria, they include some hyperthermophiles, many thermophiles, and still more mesophiles. Taken together, the population profile suggests that eucarya have always been primarily cool beasts, a few of which have adapted to hot conditions, whereas archaea and bacteria started out preferring the heat but some have evolved down the temperature scale.

The genetic makeup of archaea points compellingly to their being ancient relics from the scalding depths. If that is true, then these microbes will provide a snapshot of what life, and Planet Earth, were like in the far past. Or the argument can be inverted: if the archaean microbes' lifestyle matches what we know about the far past, this bolsters the theory that these organisms are miniature time capsules.

Let them eat rock

However justified our fears may be about endangered species and loss of biodiversity, life as a whole has a tight grip on our planet. Over time, the Earth has been molded and adapted to suit the requirements of biology. Even the impact of a large asteroid, though creating havoc and destroying many species in one blow, hasn't eliminated the ecosphere as a whole for at least 3.5 billion years. Life on Earth today enjoys a robustness and a diversity that guarantee it would survive, in some form, all but the most violent calamity.

The situation 3.9 billion years ago was quite another matter. The fate of any micro-organisms struggling to gain a toehold must have been very much in the balance. Before they could diversify to offer protection against the unexpected, the microbes had to run the gauntlet of many perils besides giant asteroids. The most pressing problem they faced was a food crisis—or, more strictly, an energy crisis. Without abundant existing life, there was nothing organic to eat, so they had to acquire their energy from somewhere else. The two possible sources were sunlight and chemicals. Given that photosynthesis is a complicated process, chemotrophy seems the more likely method.

Delving into history, we can discern the first clue to chemotrophy from the work of one Sergei Winogradsky, a Russian bacteriologist who in the 1880s studied the filamentary bacteria that inhabit sulfurous springs. He found that the genus *Beggiatoa* eats sulfur. Actually, this happens only as a desperate measure. Winogradsky was able to culture the bacteria using a medium of hydrogen sulfide dissolved in water. Although it is a deadly poison to most organisms, this chemical was clearly the sustenance of choice for *Beggiatoa*. Raw sulfur was also acceptable, but only as a starvation ration.

Winogradsky's discovery was a revelation. Until then, biologists had assumed that all organisms either ate bits of other organisms or got their energy from photosynthesis. Yet here was a microbe happily making a living from a diet of hydrogen sulfide or sulfur—both entirely inorganic chemicals. It was Winogradsky who coined the name "autotrophs" for organisms that obtain their energy from inorganic sources. Ironically, it turned out that *Beggiatoa* are not true autotrophs after all, but Winogradsky was certainly on the right lines, and many chemotrophic microbes have since been discovered. One such is the remarkable *Thiobacillus thio-oxidans*, the sulfur-guzzling bacterium that attacks sewers.

Chemotrophs make biomass from carbon dioxide, which has always been readily available on Earth, either as a gas or dissolved in water. Energy can be supplied by a variety of chemical reactions. One of these is the oxidation of sulfur or hydrogen sulfide, which is popular with surface-dwelling bacteria that have access to oxygen from the air. Of more interest to us here are the anaerobes, the oxygen haters, because free oxygen was absent on the early Earth. Among the fifty or so identified species of hyperthermophiles, the organisms with the highest growth temperatures include *Pyrodictium* and *Pyrobaculum*. They have no truck with oxygen at all, which accords well with the theory that these heat-loving archaea are living fossils from an oxygen-free era of long ago. These superbugs obtain their energy from sulfur by combining it with hydrogen to make hydrogen sulfide.

Sulfur is scattered widely among important biomolecules, a minor but important chemical in extant life. Sulfur-metabolizing bacteria include some of the most ancient hyperthermophiles. This points to a key role for sulfur in the formation of life. The old name for sulfur is "brimstone," a devilish substance associated with fiery volcanoes and hell. It was a common element on the primeval Earth, especially in the form of hydrogen sulfide. The central place of sulfur in the story of life's origins is an amusing irony. Not only was the real Eden most likely a Hadean inferno, it may also turn out that life was created from brimstone!

Iron is another element important for life. It is often found in combination with sulfur as the mineral pyrite, commonly known as fool's gold. Pyrite has been suggested as the chief catalyst for biogenesis by the German chemist Gunter Wächterhäuser;[20] iron-sulfide membranes also form the setting for Mike Russell's theory of the origin of life, which I described in chapter 5. Pyrite is still a food source for the chemotroph *Thiobacillus ferro-oxidans*, which obtains energy from the oxidation of both the iron and the sulfur components. Incidentally, mining engineers are well aware of the activities of this busy organism. The ferric iron produced as a waste product liberates more iron and sulfur from the pyrite, creating a runaway cycle of reactions. When large amounts of pyrite occur in ore dumps, mines, or coal seams, this process can corrode machinery and create a serious pollution problem known as acid mine drainage. *Thiobacillus ferrooxidans* can also digest other mineral sulfides, such as copper, tin, and even uranium, and has been exploited commercially in mineral refining. Another iron-eating chemotroph that advertises its presence is *Gallionella*, an inhabitant of iron-rich streams. It converts soluble ferrous salts into the insoluble ferric state, leading to the distinctive rust coloration of the water. Sulfur and iron might well have been the chief midwives at life's birth in the crust of the Earth, and they continue to offer rich pickings for micro-organisms today. Next time you see a rust-colored stream, reflect on the fact that you might be witnessing a process directly related to life's origin.

Many other chemotrophic pathways are exploited by microbes. The remarkable archaea fall naturally into three groups: thermophiles, halophiles (salt lovers), and methanogens. The latter get their energy by making methane, a very basic form of metabolism that is still widespread in the microbial world. If you are an accomplished chemist, you can make methane directly from hydrogen and carbon dioxide. That is what *Methanothermus*—a rod-shaped microbe that inhabits the hot springs of Iceland—does. Recently, Todd Stevens and Jim McKinley of the Pacific Northwest Laboratory in Richland, Washington, stumbled across microbes doing something like this deep under the ground, during a drilling project in the Columbia River region.[21] They were alerted to the presence of subterranean methanogens after an explosion occurred as they were boring through a deep basalt layer. On investigation, Stevens and his colleagues found that the deep rocks were giving off hydrogen. Hydrogen gas is highly explosive in air, and it came as a surprise to me to learn that it is found occurring naturally anywhere on Earth these days. Apparently, various chemical processes will produce it, such as the infusion of acidic water into iron-rich silicates. Amazingly, there are certain locations in Oman, California, and Japan where very high concentrations of hydrogen seep to the surface.

Hydrogen gas is a welcome source of energy for methanogens, which combine it with dissolved carbon dioxide and make biomass as they go. In doing so, they may be enacting the oldest form of metabolism. These are examples of chemotrophs that are truly independent of surface life, and do not rely, even indirectly, on the products of photosynthesis. As such, they could support a food-and-life chain that would flourish in total darkness, deep beneath the surface. This conjecture is not an idle speculation. Stevens and McKinley managed to culture bacteria from their borehole, and found that some of the microbes live parasitically off the organic material produced *ab initio* by the others. The scientists believe that a complex ecosystem resides within the Columbia River basalt, and they coined the evocative nickname SLIME (for Subsurface Lithoautotrophic Microbial Ecosystem) to describe it. Almost certainly there are SLIMEs in many other locations awaiting discovery.

The methanogens occupy one of the deepest branches of the archaean tree, and so by implication they are among the earliest life forms. One methanogen, called *Methanopyrus*, also has one of the highest growth temperatures (110 degrees Celsius) and contains a peculiar membrane chemical that looks like a precursor of the lipid membranes found in most archaea. These features suggest that *Methanopyrus* may be among the most primitive types of organism found so far.

The problem about reconstructing the microbial base of the tree of life is that we have no idea what organisms may remain undiscovered at this time, perhaps in some obscure SLIME. Microbiologists are finding not only new species all the time, but occasionally entire new kingdoms. The Obsidian Pool in Yellowstone National Park recently yielded two hitherto unknown archaea that occupy a distinctive deep branch situated between the eucarya and the main group of archaean species. Sequencing by Susan Barns, Norman Pace, and their colleagues indicates the existence of a separate grouping that may represent the most primitive microbial kingdom known.

Of course, no extant organism will be an exact clone of its ancient ancestors. Some evolutionary drift is inevitable over such great durations of time. Nevertheless, we can try to guess which known microbe might bear the closest resemblance to the universal ancestor. A likely candidate is the sulfur reducer *Pyrodictium*. It too thrives at 110 degrees Celsius, suggesting a deep thermophilic ancestry. It lives in colonies that form curious networks of cocci connected by tiny filamentary tubules. It is fascinating to wonder whether our distant forebears inhabited such a tangled web in some torrid subsurface niche, almost four billion years ago.

The rest is history

To summarize: The record of the genes suggests that the universal ancestor lived deep beneath the Earth's surface, at a temperature well above a hundred degrees Celsius, and probably ate sulfur. However, it is clear that this little creature was already a sophisticated life form with complex features like coded protein synthesis. As I have stressed before, the universal ancestor was not the first living thing. A long evolutionary history must have preceded it. We know almost nothing about the circumstances that connected the first living thing to the universal ancestor.

It is tempting to speculate that life actually began in a geothermally heated, mineral-rich, subsurface niche, and evolved *in situ* as far as the universal ancestor, before radiating out across the planet. However, we do not know if this was the case. Life may have started in a completely different environment altogether, and invaded the hot subsurface region at a later date. In chapter 6 I discussed the work of Norman Sleep and others suggesting that the Earth's surface suffered episodic sterilization by the rock vapor from massive cosmic impacts. According to this "frustration" theory, life kept getting wiped out, only to emerge again, phoenixlike, from the ashes. As the bombardment tapered off, the surface would still get seared from time to time, but refuges would now exist in the deep rock strata. Because these deep rocks were geothermally hot, they could offer a home only to hyperthermophiles. Even mesophiles would have died there. It is, then, no surprise that the universal ancestor was a hyperthermophile; such were the only organisms whose comfort zone lay beyond the reach of the cosmic-impact heat pulses. Any cold-loving surface microbes that may have preceded the thermophiles would have been cooked by impacts, their particular branches of the tree of life abruptly truncated. If this scenario is correct, then the position of hyperthermophiles near the base of the known tree of life does not necessarily indicate that life began hot and deep, only that life on Earth had to pass through a temperature "bottleneck" created by the meteoric barrage.[22]

A pointer to an earlier phase of life comes from the discovery of autotrophic bacteria that can synthesize biomass not only from scratch using carbon dioxide, but also by using more complex organic substances such as acetic acid. The organisms that do this have been dubbed mixotrophs, and they use as an energy source either light, as in the case of green sulfur bacteria, or a chemical reaction such as the oxidation of sulfur or hydrogen. If the surface of the primeval Earth was coated in organic substances from cosmic impacts, this would have provided a ready supply of raw materials. Perhaps the very first organisms were surface-dwelling mixotrophs, and their metabolic habits live on in a handful of microbes. Of course, by the same token, life may have started in the comets themselves, an idea I shall come back to in chapter 9.

Though we can't pinpoint where life ultimately began, it seems increasingly likely that, after the bombardment abated, life was confined to locations on or beneath the seabed, either near volcanic vents, or inside off-ridge hydrothermal systems. Once life had established itself securely in such a place, the way then lay open for proliferation and diversification. Granted that what follows is mainly guesswork, I think the subsequent story would have gone something like this: The earliest microbes were hyperthermophiles, relishing temperatures of 100–150 degrees Celsius. They dwelt at least a kilometer beneath the surface, possibly on the seabed, but more likely in the porous rock beneath it. Immersed in superheated water replete with minerals, they greedily ingested and processed iron, sulfur, hy-

drogen, and other readily available substances, releasing energy
from primitive and rather inefficient chemical cycles. These early
cells were crude rock-eaters. Neither light nor oxygen played a role
in their metabolism. Nor did they require organic material; they
made what they needed directly, from the rocks, and carbon dioxide
dissolved in the water.

The first microbial colony had the whole world at its disposal,
and a plentiful supply of materials and energy. It would have spread
with amazing speed. The potential for microbes to multiply explo-
sively fast guaranteed that they would swiftly invade every accessi-
ble niche. With no resident competition, they could rapidly inherit
the Earth. However, given the population explosion, the colony
would soon have reached the limits of its habitat. Barred from going
deeper by the rising temperatures, and unable to reproduce in the
cooler surface strata, the microbes could spread only horizontally
along the mid-ocean volcanic ridges, and laterally through the
ocean-floor basalt.

At some stage, perhaps 3.8 billion years ago, the first great evolutionary fork was reached, when a group of microbes suddenly
found themselves cut off from their warm and snug haven due to
some geological upheaval like an earthquake or a volcanic eruption.
Isolated from the main colony, and stranded in a cooler region, almost all the microbes became dormant or simply died, their membranes too rigid at these lower temperatures to enable their
metabolism to function. However, a lucky mutant that accidentally
had a more flexible membrane survived and multiplied. By making
the transition to cooler conditions, the mutant microbe paved the
way for access to the uninhabited surface of the planet. Meanwhile,
for the members of the original colony, confined comfortably in
their subterranean realm, life has continued much the same up to
the present day.

A key early development was a switch by some organisms from
chemicals to light as a source of energy, by which stage life must
have spread as far as the surface. The first such "phototrophs" probably did not use modern chlorophyll photosynthesis, but some more

elementary process. Certain Dead Sea archaea still use a rather primitive form of photosynthesis based on a red substance related to vitamin A. Capturing sunlight began in earnest with the bacteria, which discovered a way to pluck electrons from minerals, boost them with solar photons, and use the stored energy to make organic material. A later refinement freed up the dependence on minerals, by enabling bacteria to strip electrons out of water, releasing oxygen as a consequence. The crucial component in this ingenious process was chlorophyll, the substance that makes plants green. With the only requirements being water, carbon dioxide, and light, the way was open for the greening of the planet.

Still unanswered is how and when the three great domains arose: archaea, bacteria, and eucarya. It seems probable that the great split in the tree of life between archaea and bacteria occurred before the invention of photosynthesis, perhaps as early as 3.9 billion or 4 billion years ago—well inside the era of heavy bombardment. The evidence points to the archaea's being the oldest and most primitive organisms, with bacteria arising somewhat later. So deep was the cleft between the archaea and the bacteria that they have never really been rivals; they still occupy different niches after several billion years of evolution. Finally, the deep rift that produced the eucarya domain probably occurred when conditions were somewhat cooler. For some reason, perhaps by being exposed to the challenges of a less stable environment, the lower-temperature eukaryotes evolved at a much faster rate. The subsequent flowering of life, its diversification into many species, and the huge rise in biological complexity stemmed directly from the branching away of eucarya on the tree of life. Without this momentous step, it is unlikely that we—or any other sentient beings—would exist on Earth today to reflect on the significance of it all.

CHAPTER 8

Mars: Red and Dead?

*That Mars is inhabited by beings of some sort or
other is as certain as it is uncertain what those beings
may be.*

PERCIVAL LOWELL, 1906[1]

ON AUGUST 7, 1996, President Bill Clinton faced the world's press and announced in dramatic terms that NASA had evidence of life on Mars. Clinton was referring to the discovery of a Martian meteorite found in Antarctica in 1984, containing what could be signs of life. He went on to remark that, if this stunning discovery held up, it would serve to redefine mankind's relationship with the cosmos.

The possibility of life on Mars has a long history. In the seventeenth and eighteenth centuries, scientists, philosophers, and even theologians speculated freely about Martians, Venusians, and other extraterrestrial beings. By the end of the nineteenth century, however, astronomers had become much more skeptical about the prospects for life on other planets. Nevertheless, in 1877, the Italian astronomer Giovanni Schiaparelli reported that he had seen a pattern of straight lines on the Martian surface. He used the Italian word *canali*, meaning "channels," to describe these features. In the United States, Percival Lowell and others seized upon the idea and claimed that Schiaparelli's *canali* were in fact artificial canals. Lowell believed that the Martians had built the canals to irrigate the parched terrain, using meltwater from the polar caps. He built an observatory in Flagstaff, Arizona, dedicated to mapping the canal network. In Lowell's eyes, Mars was a dying, drying planet. It followed that any intelligent Martians might well be desperate, and driven to construct a massive irrigation system. The theme of the dying planet, and of envious Martian eyes turned upon our own green and pleasant world, was brilliantly exploited by H. G. Wells in his famous novel *War of the Worlds*, published in 1898.

Few astronomers went along with Lowell's idea of canal-building Martians, and as observations improved so the chances of finding life on Mars dwindled. Nevertheless, some scientists remained convinced that a primitive form of vegetation, perhaps a type of lichen, grew there. They pointed to seasonal changes in color as evidence. But even this possibility was abandoned with the advent of the Space Age. Probes sent to the red planet found no signs of life, let alone any canals.

In 1977, NASA finally put the matter to the test directly, by landing two *Viking* spacecraft on the Martian surface. The craft were specifically designed to seek out life. By this stage, few people hoped for more than some microbes in the Martian soil. The data sent back by *Viking* confirmed the skeptics' opinion. The soil tests failed to find any convincing evidence for Martian microbes. To the disappointment of many, the red planet was pronounced a dead planet.

For twenty years after *Viking*, the idea of life on Mars was largely dismissed as science fiction. And so it might have remained were it not for a series of astonishing discoveries—not on Mars, but right here on Earth. These discoveries have put an entirely new complexion on the subject. It now appears that scientists may have been a bit too hasty in writing off Mars as an abode for life.

A bad place for a vacation

Visually, Mars is a spectacular planet. It shines with a rich red hue in the night sky, earning it the ancient name of the god of war. Telescopes reveal white polar caps and great dusky patches. Occasionally dust storms shroud the entire planet. Close-up pictures from space probes show a surface peppered with craters and riven by giant canyons and valleys. Huge extinct volcanoes dot the landscape. On the ground, the terrain resembles the most desolate parts of the Australian desert: ocher-red soil strewn with boulders, and fine sand blown into dunes. The entire vista is bathed in a watery sunlight beneath an orange sky.

From the point of view of life, Mars presents every conceivable hazard. The temperature is nearly always below freezing, and can dip as low as –140 degrees Celsius. The atmosphere, consisting mainly of carbon dioxide with mere traces of oxygen and nitrogen, is pitifully thin. At 7.5 millibars, the air pressure is no higher than on Earth at 35,000 meters—considered to be the edge of space. There is no protective ozone layer, so the surface is drenched with deadly ultraviolet from the Sun. The soil is corrosively oxidizing, and so dry it makes the Sahara seem like a swamp. In fact, if the total water-vapor content of the Martian atmosphere were dumped on the ground, it would barely dampen the soil. The dryness makes the sandstorms quite fearsome. When the wind whips up, sometimes reaching 650 kilometers per hour, the dust can soar to a height of 50 kilometers. All in all, Mars would not be a nice place to be stranded.

The root cause of the planet's uncongenial conditions can be traced back to its small size. It is about half the diameter of Earth, with only 38 percent of Earth's surface gravity. As a result, most of its atmosphere has leaked away into space. The low pressure prevents liquid water from existing on the surface even above the freezing point; pour out a cup of tea on Mars and it would immediately evaporate. The thin atmosphere also means that greenhouse warming is feeble. The cold is exacerbated by the planet's distance from the Sun, which averages 228 million kilometers—about 50 percent more than Earth's.

You might think that any search for life on such a frigid and desiccated planet would be a complete waste of time. However, even in the 1970s, when the *Viking* mission was being planned, scientists knew that some bacteria can survive in cold, dry conditions like Antarctica, so they devised a suite of onboard experiments designed to seek out microbial life in the Martian soil. Robot arms were installed on the spacecraft that could reach out, scoop up some dirt, and deposit it in a mini-laboratory for analysis.

Three experiments were performed on each of the two spacecraft. The first was called the gas-exchange experiment. It consisted of pouring a nutritious broth onto soil samples and monitoring the release of any gases. Before the nutrients were added, the samples were exposed to water vapor. To the surprise of the scientists, this initial step provoked a vigorous response, in which copious amounts of oxygen were given off, along with some nitrogen and carbon dioxide. Similar results were obtained from soil exposed to sunlight and concealed beneath rocks. When the soil was preheated to 145 degrees Celsius, thought at the time to be enough to kill all terrestrial microbes, the oxygen release seemed to be affected, although some doubt was cast on the reliability of this result. When the broth was eventually added, further complicated gas exchanges took place, but no systematic pattern was discernible. Certainly the Martian soil didn't behave at all like terrestrial soil. The mission scientists were a bit nonplussed, and concluded that the surface of Mars must be chemically highly potent, so that the simple expedient of adding water had the effect of making the soil fizz. Microbes weren't needed to explain what happened—although, to be fair, the gas-exchange experiment didn't rule them out either. At best, the results were ambiguous.

Next on the list was the so-called labeled-release experiment. This also involved adding broth to the soil, but the mixture was different. Crucially, it contained a radioactive-carbon tracer, and the gases released were tested for signs of radioactivity. The assumption was that any Martian organisms that processed the carbon and liberated carbon dioxide would generate some radioactive gas. This could be detected with great sensitivity. In the event, the labeled-release experiment gave a positive result. Moreover, when the soil was strongly heated, the result turned negative, exactly as expected if micro-organisms had been at work.

Third came the carbon-assimilation experiment. In a sense, this was the reverse of the labeled-release experiment. Soil samples were exposed to an atmosphere of radioactive carbon dioxide and illuminated by a strong light source to simulate the Sun. The object was to see if any carbon was taken up by Martian organisms as part of their growth process, in the same way as terrestrial plants use up carbon dioxide. This also gave a positive result in several of the runs. Heating the sample to 170 degrees Celsius diminished, but did not entirely eliminate, the response.

Taken at face value, the *Viking* experiments could be seen as offering some evidence for microbial life in the Martian soil. However, NASA scientists were almost unanimous in drawing the opposite conclusion. The behavior of the soil samples was sufficiently complicated and unexpected to cast doubt on a straightforward biological interpretation, and opinion inclined more towards the belief that unusual soil chemistry, probably involving strong oxidation, was responsible. This conclusion was supported by the fact that *Viking* found no trace of organics in the Martian soil, which is odd: even if there is no life on the surface of Mars, some organic material must have been delivered there from space. The explanation seems to lie with the fierce ultraviolet radiation and oxidizing soil, which would tend to break up any organic molecules strewn on the surface.

Taken together, the *Viking* experiments fall short of clear-cut evidence for life on Mars, and the official conclusion of the mission was that Mars is a lifeless planet. However, we must always remember the dictum that the absence of evidence is not the same as evidence of absence. There are many reasons why *Viking* may have failed to detect life on Mars, other than the obvious one that there isn't any:

- The experiments may have been testing for the wrong sort of life. They were designed to respond to terrestrial organisms. Martian life may be based on an entirely different biochemistry, or temperature range. The conditions in the Viking mini-lab may have been comfy for terrestrial microbes but lethal for Martians.

- The crucial experiments may not have been sensitive enough to detect a relatively low density of Martian micro-organisms in the soil. (As many as a million microbes per gram would have gone unnoticed.)

- The Martian topsoil is sterile, but life may still exist deep in the cracks of rocks, which afford some protection from the harsh conditions.

- The experiments may have been suitable, but not the landing site. Perhaps life exists on the surface of Mars in certain favored niches, away from the two landing zones.

- Life may exist on Mars, but not on the surface. Suitable habitats might be situated beneath the polar caps or deep underground—a possibility I shall return to shortly.

Even if all these points are discounted, Mars could still be of major interest to biologists, for a simple reason. Today the red planet may present a bleak picture, but it was not always a frozen wasteland. There is abundant evidence that in the remote past Mars was warm and wet and Earth-like, and much more hospitable for life. Whether or not Mars is today a totally dead planet, there is still a good chance that life may once have flourished there.

Flood

You can easily tell that Mars was once more favorable for life by glancing at the pictures taken by the *Mariner* and *Viking* space probes. One distinctive feature leaps out of the survey photographs: river valleys. There, among the tangled mountain uplands, cutting swathes across sandy plains, carving deep into hillsides, spilling from the rims of craters, are easily recognizable channels sculpted by running water. They come complete with tributaries and deltas and flood plains. These watercourses, I might add, bear no resemblance to Lowell's famous straight-line canals; instead, they are dendritic and sinuous, like rivers on Earth, and undeniably natural rather than artificial.

Unfortunately, no trace of water remains in Mars' ancient riverbeds; they have long since dried up. But we can be confident that these valleys are indeed fluvial, for they display all the familiar features of terrestrial rivers, such as cataracts, streamlined bank sides, and teardrop-shaped islands, where silt has been swept along and deposited by the current. There can be no doubt: water once flowed freely on Mars. But where did it come from? Where did it go? Were Martian watercourses conventional rivers fed by rain and melting snow, or sourced by underground springs and aquifers? Did the rivers discharge into lakes and seas, or simply run into the sand? Above all, how long ago did these river valleys form?

Scientists have spent years poring over the survey photographs, squeezing information out of every minute detail, in a valiant attempt to address these queries. Even from a cursory inspection it soon becomes clear that many of the bigger channels are not so much riverbeds as flood plains, scoured by raging torrents following the sudden release of huge bodies of water. The shape provides the clue. A cataclysmic outflow typically creates a channel that starts abruptly, full-sized and deep, with few side channels. A river, by contrast, begins as a trickle and builds up in both size and depth as tributaries feed into it.

When numbers are put in, the scale of the Martian floods is staggering. Channels vary from a few tens of kilometers across in the highlands to basins of eroded land hundreds of kilometers wide, where water once surged across open plains. In full spate, the flow rate along the larger channels would have been prodigious, equal to ten thousand Amazons. The largest known cataclysmic flood on Earth overwhelmed the Columbia River in Washington State some twelve thousand years ago. In that episode, a volume of water comparable to Lake Michigan flowed away in just two days. Martian floods were up to three hundred times more powerful than this.

The precise cause of the huge inundations on Mars remains in dispute. Almost certainly they were not due to heavy rainfall. What seems to have happened is that large pent-up bodies of water suddenly broke free. Most likely this was liquid restrained by an ice dam that melted and collapsed. Another possibility is that underground water burst through a permafrost seal on to the surface, like a colossal fountain. Such an eruption might have occurred when a meteorite pierced the crust, or from volcanic melting, or simply as a result of a buildup of hydrostatic pressure.

Not all the dried-up watercourses on Mars are catastrophic outflow channels. In the older terrain of the southern highlands, there are many features that look much more like conventional river systems, with long narrow valleys, delicate gullies, and slow erosion of the land. These riverbeds are tens of kilometers long and up to three kilometers wide, and possess tributary networks similar to those on Earth. Opinions differ as to how these valley networks formed. The simple picture of rain or snow producing runoff that flows downhill and slowly erodes the valley floor doesn't match the facts too well. Certainly it couldn't happen today, because the water in the small streams would evaporate or freeze solid before it could discharge into the main river. But even if conditions were once conducive to liquid water, the shapes of the valleys don't correspond well with runoff erosion.

Another process that forms valleys here on Earth is called groundwater sapping. You can see it at work on a small scale on sandy beaches, when a spring bubbles up into the sand and the water flows out to sea. As the system evolves, the head of the stream works its way back up the sand, cutting a wide path in the upstream direction as it goes. Many Martian valleys look as if they have formed in this manner.

Michael Carr of the U.S. Geological Survey is a leading expert on Martian water. He believes that very little fluid actually flowed on the surface to create the valleys. He points to the flat floors and steep sides of the channels, suggestive of a form of subsidence. Carr thinks that most of the water seeped under the ground, steadily undermining the land and causing it to sag or waste, rather than running over the surface and wearing away the material. The lubricating effect of subsurface flow can cause loose material to slide downhill, creating a gully even in the absence of surface erosion. Carr thinks that, rather than a rainfall cycle operating, some sort of geothermally powered convection cycle must have been at work, to return the discharged water repeatedly to underground aquifers. The overall picture of water on Mars is therefore one of slow formation of valleys by steady surface or subsurface flow, plus the occasional sudden and catastrophic flood.

When I was a teenager, I got mischievous enjoyment from arguing with Jehovah's Witnesses. My favorite question concerned Noah's flood: where did all the water go? We can ask the same of the Martian floods. The simple answer is: into the ground. Like the Earth and Moon, Mars suffered intense cosmic bombardment during its first seven hundred million years. The tumult churned up so much material that it covered the whole planet with rubble, known as regolith, to a depth of several kilometers. Because Mars is much smaller than Earth, it lacks a large molten core to drive tectonic activity, so this fragmented surface has not been widely reprocessed. A porous regolith therefore remains like a vast sponge, capable of sequestering a huge amount of liquid. Thus, even though the surface is now extremely dry, Mars may still have extensive reserves of water concealed beneath the ground, in the form of permafrost or, many kilometers down, as trapped liquid. Estimates vary, but it seems likely that if all this water were released at once and deposited on the surface it would form a planet-wide ocean with a depth of at least a kilometer.

Some Mars-watchers think the red planet once had extensive seas and lakes in spite of the porous surface. Traces of ancient lake sediments, layered and very thick, exist in many deep canyons; the mottled appearance of some low-lying northern plains also suggests widespread ponding. Evidence for a large sea is more controversial, but a possible ocean boundary can be traced around the northern lowland plains, where large outflow channels from the cratered uplands discharged their water in balmier days. The putative shoreline includes eroded cliffs, wave terraces, and cusps. Dubbed Oceanus Borealis, this Martian sea may have covered a third of the planet.

Complementary to evidence for an ocean are strong signs that the southern hemisphere of the planet has been subjected to large-scale glaciation. Today Mars has a thin northern polar cap containing water ice mixed with dry ice (frozen carbon dioxide), and a more substantial southern polar cap of predominantly dry ice. The caps wax and wane with the seasons; the northern cap can disappear completely. But long ago a thick water-ice sheet extended from the South Pole as far as 33 degrees latitude. The source of all this ice may have been evaporation from Oceanus Borealis.

Over the eons, Mars has been gradually drying out as water vapor is lost to space because of the low gravity. An amount of water equivalent to a global depth of seventy meters could have been relinquished that way. More serious is the cold. As the temperature slumped, conditions became unsuitable for liquid water, and most of the Martian seas became incorporated in the permafrost. Ancient discharge lakes would probably have frozen solid at high latitudes, and their remnants may still be there, obscured beneath layers of dust and rock.

Though scientists are divided about the details of water on Mars, they agree that most hydrological activity happened a very long time in the past. If ever there were serene meandering rivers, or churning oceans, they probably dried up at least three and a half billion years ago. However, the degeneration of the climate may not have been a one-way street. The slow desiccation could have been punctuated by short warmer episodes, when water once again ran freely. Evidence comes from the fact that some Martian valleys formed quite late on. Also, some of the larger outflow channels have clearly been cut several times, indicating a sequence of flooding episodes. All this suggests that from time to time Mars returned, perhaps only briefly, to warm and wet conditions for some reason. There may then have been extensive recycling of water through the ground and atmosphere. But with each cycle of flooding and glaciation, more water disappeared. Although some rivers may have run on Mars as recently as a few hundred million years ago, they were feeble in comparison with the ancient floods, and would have had little effect on the Martian climate.

The Martian greenhouse

Martian rivers offer clear-cut evidence that the planet was once warmer and wetter. But how was this possible? At first sight, there is good reason to believe that Mars should have been even colder in the past than it is today. This has to do with the so-called dim-young-Sun problem. As the Sun ages, it grows slowly brighter, due to changes in its chemical makeup. Four billion years ago, it would have been 30 percent dimmer than it is today, drastically reducing its heating effect on distant Mars. Set against this is geothermal warming, produced by radioactivity and the stored heat from the planet's formation, both of which were much higher in the past. However, geothermal heat flow alone wouldn't compensate for the dim young Sun, and other reasons for a milder climate must be found.

The easiest way to make a planet warmer is by using the green-house effect. Greenhouse gases such as carbon dioxide act like a blanket, trapping the Sun's heat near the planet's surface. Today the Martian atmosphere is too thin to produce much greenhouse warming, but it would certainly have been a lot thicker during the first billion years. As with Earth, Mars acquired a dense early atmosphere both from outgassing of the planet and from the delivery of volatile substances by comets, asteroids, and icy planetesimals. Abundant CO_2 would have boosted the temperature dramatically.

Although scientists guess that Mars must have had far more CO_2 in the past, putting a figure to it isn't easy. It first has to be determined where the CO_2 has gone. Very likely most of it was lost to space as a result of massive cosmic impacts. As I explained in chapter 6, the collision of large comets with the planets causes impact erosion, stripping away the atmosphere. In the case of Mars, the end result was seriously thin air, but during the bombardment period itself the pressure would have fluctuated wildly. Calculations suggest that Mars lost 99 percent of its atmosphere from impacts during the first seven hundred million years, and a further 90 percent of that thereafter due to a variety of processes. If these numbers are right, they imply that at times Mars may once have had an atmospheric pressure a thousandfold higher than now—enough to lift the temperature above freezing and even support an extensive ocean.

There isn't much doubt that Mars once had a thick atmosphere, because the walls of the older impact craters have been extensively weathered. Craters smaller than fifteen kilometers across have been completely annihilated. By contrast, later craters are hardly eroded at all. Dating the change, investigators think that the atmosphere thinned dramatically not long after the end of the late heavy bombardment, 3.8 billion years ago. Most of the catastrophic floods seem to have occurred before or at about that time, because the discharge channels are embellished by a lot of well-preserved small craters. It is the lack of erosion for most of Mars' history that has kept its extremely ancient watercourses in virgin condition. On Earth, no river valley would survive for billions of years.

Once the bombardment declined, Mars' carbon dioxide continued to leak away, from a variety of causes. Some escaped into space, some dissolved in the water or became absorbed into the regolith, and a lot may have become incorporated into carbonates or other minerals in the rocks. Without some compensatory process, the CO_2 would have been gobbled up in pretty short order. Probably geothermal heating reversed some of these processes and returned part of the CO_2 to the atmosphere. For some hundreds of millions of years, there may have been a moderately high atmospheric pressure and associated greenhouse warming. Eventually, however, the geothermal heat faded, the CO_2 recycling faltered, and the atmospheric pressure started to plummet, producing the freeze-dried desert we see on Mars today.

The fact that some river valleys seem to have been cut relatively recently suggests occasional episodes of warming. A possible explanation comes from feedback processes. If local geothermal heating or an outburst of volcanism were suddenly to release large amounts of water to the surface, then a lot of dissolved carbon dioxide would be given off with it. This in turn would elevate the temperature, melting more water and liberating more CO_2. As the meltwater inundated the frozen lowlands, it would warm the regolith, yielding yet more CO_2. All in all, enough carbon dioxide could be released from the planet in this runaway fashion to create temporarily a denser atmosphere with pronounced greenhouse warming.

Another wild card concerns the motion of the planet. Mars has a rather eccentric orbit, and no moon to stabilize its spin axis. There would have been times when favorable combinations of the spin and orbital motion led to considerably enhanced solar warming. On occasions, the spin axis might have tipped right over, so that the poles received more sunlight than the equatorial regions. This would have melted the polar caps and produced an escalating greenhouse effect. On balance, repeated episodes of flooding, ocean formation, and glaciation followed by long periods of inactivity seem more likely than simple uninterrupted cooling.

Concerning the possibility of life, the fact that Mars was warm and wet between 3.8 and 3.5 billion years ago is highly significant, for it means that Mars resembled Earth at a time when life existed here. This has led some scientists to conclude that Mars would have been a suitable abode for life at that time too. On its own, however, the presence of liquid water is only part of the story. What makes the prospects for life seem so good is that Mars had not only liquid water but also volcanoes.

Was there life on Mars?

The Martian mountain of Olympus Mons towers 27 kilometers above the Tharsis massif and is fully 550 kilometers across. Measure for measure, it is the biggest mountain of its type in the solar system, equivalent to stacking up seven Mount Everests on Earth. The significance of Olympus Mons lies not with its size, however, but with the fact that it is a volcano. Where volcanoes and water come together, hot springs can result—hydrothermal systems like those on Earth that possibly played host to the first organisms. Did microbial life also flourish on Mars 3.8 billion years ago, in some gurgling fountain on the slopes of Olympus Mons perhaps, or deep in the porous rocks beneath a long-vanished Martian sea?

Four billion years ago, Mars still glowed from the heat of its formation. Radioactivity warmed the crust. Cosmic impacts melted the surface. As the planet struggled to divest itself of this primeval heat, it spewed forth lava from volcanoes on a massive scale, creating immense plains of molten rock like the maria of the Moon. As the crust slowly cooled, this volcanism steadily declined; by the time the heavy bombardment ceased, it was largely confined to three main regions: Tharsis, Elysium, and Hellas. If there are live volcanoes on Mars today, they aren't showing any signs of activity.[2] However, there have been eruptions throughout Martian history: for example, around Olympus Mons within the last one and a half billion years, and near Alba Patera as recently as five hundred million years ago. Since it is unlikely that Mars would be volcanically active for four billion years only to cease relatively recently, it seems reasonable to conclude that some hot spots still exist, probably deep underground.

In the remote past there must have been ample opportunity for hot springs to form around thermal vents, given the abundance of water on the planet. There is clear evidence of the interaction of water and volcanoes in the survey photographs. Many of the floods were likely triggered by lava melting the permafrost and ground ice, and some watercourses are clearly seen to emerge from beneath lava flows. Outflow channels also cluster around the highly volcanic Tharsis region. Elsewhere, dense valley networks decorate the flanks of volcanoes. There are flat-topped hills that look like table mountains in Iceland where lava has oozed beneath ice. Characteristically shaped ridges in Elysium also bear the hallmarks of a lava-and-ice combination. All this adds up to strong circumstantial evidence of hydrothermal systems on ancient Mars, though specific mineral deposits—which would be a clear giveaway sign—have yet to be detected.

While they await new Mars missions, NASA scientists have been busily earmarking spots on the planet's surface where hydrothermal activity could have occurred. The side of the Hadriaca Pladera volcano seems a good place. Here are to be found many tangled river valleys flowing away from the rim of the ancient caldera, cut through by a spectacular flood channel that emerges abruptly partway up the slope. Another volcano, Apollinaris Patera, features an odd-looking bright patch near the caldera rim, which could be a hot-spring mineral deposit. A similar volcano in the highly cratered area known as Terra Cimmeria has highly eroded slopes and is situated at the commencement of a large watercourse.

Many river valleys on Mars occur in chaotic terrain where great blocks of rock lie in tangled masses. This topography is thought to have formed when molten rock intruded into ground ice. As the ice melted, the water flowed away, causing the land to collapse in a haphazard manner. Such areas would be a prime site for shallow hydrothermal systems to arise.

If life did take up residence in a hot spring, it may have left fossilized remains. Martian fossils are likely to have withstood the ravages of time better than their terrestrial counterparts because of the relative lack of weathering. Future landing missions could seek out likely-looking samples for return to Earth. Other potential fossil sites include river valleys, where floods may have swept tiny Martian organisms into stagnant pools, and the huge rift valley, Valles Marineris, where deep strata have been exposed. Dried-up lake beds, where microbes could have become deposited in sediment, are also of interest. The crater known as Gusev looks to be a promising candidate, because a large river once flowed into it. There must have been a deep lake there long ago, with lots of sediment on the bottom.

202 ° **THE FIFTH MIRACLE**

The first small step in following up these pointers came in July 1997, when the *Pathfinder* mission successfully deposited the first spacecraft on Mars since *Viking*. With its little rover vehicle Sojourner, *Pathfinder* beamed back a wealth of data from the mouth of the Ares Vallis flood plain. The terrain near the spacecraft is strewn with a grab bag of rocks swept down by the torrent. This detritus might include fragments of an ancient hydrothermal system, or even fossils of deep subsurface microbes brought to the surface in the flood and conveyed downstream. Unfortunately, *Pathfinder* didn't have the capability of checking these conjectures.

In September 1997, *Mars Global Surveyor* went into orbit. It is designed to map the surface of the planet to meter-scale precision, and is yielding valuable information about the hydrological history of Mars and likely abodes for life. Recent pictures include evidence for an ancient ocean shoreline, dried-up ponds within a crater, and hints of mineral deposits associated with hydrothermal systems, all of which favor the prospects for past life. More probes are planned by NASA, the European Space Agency, Japan, and Russia, culminating in a sample return mission, maybe in 2005. Although these missions are directed largely at understanding the climate and geology of the planet, all the results will be eagerly scrutinized for clues to past life.

Is there still life on Mars?

If life did get going on the surface of Mars 3.8 billion years ago, it would have faced a desperate race against time. Hardly had the sterilizing bombardment ceased when the climate began to deteriorate. As the temperature plunged and water froze, suitable habitats would have become scarcer and scarcer. Within just a few hundred million years, any remaining organisms would in all likelihood have retreated to special refuges, such as desolate lakes protected by ice covers, or deep subsurface locations.

Is it conceivable that life is still clinging on there today? With hindsight, the sites picked for the *Viking* mission, which were chosen primarily for the ease of landing, seem about the least likely places for life. *Viking* was flown before biologists appreciated the significance of hot springs. Unfortunately, all of the hydrothermal systems on the Martian surface seem now to be extinct.

It would be a mistake, however, to dismiss Mars completely as an abode for extant life. Erupting volcanoes and vomiting vents may be largely a thing of the past, but substantial geothermal heating could still be going on deep underground. Though the permafrost extends down for several kilometers, liquid water, probably salty, could be plentiful beneath it. We know Earth's biosphere extends deep into the crust. If organisms can dwell contentedly in the subsurface zone here, they could do so on Mars too. Though Mars may lack the cornucopia of black smokers we find on our ocean floors, there is no reason why Martian microbes could not have adapted over time to that planet's more spartan conditions. On Earth, bacteria and archaea have invaded the harshest of habitats, and thrive in places that would make conditions beneath the Martian permafrost seem positively benign.

If there is life on Mars, it would probably resemble the SLIMEs found on Earth in the deep rock strata beneath the ground, supported by chemotrophs (see chapter 7).[3] Remember that chemotrophs are primary producers: they require no light, organic food, or oxygen. Their nutrients are inorganic chemicals supplied from below, such as hydrogen and hydrogen sulfide, carried vertically in the crust by water convection. The ancient metabolic processes they employ would be very suited to current Martian conditions, where sulfur and iron deposits could supply the necessary chemicals. An organism like *Methanococcus*, which converts hydrogen and carbon dioxide into methane, would probably feel at home in subsurface Mars.

How can this conjecture be tested? Getting at any live microbes under the permafrost would be tricky even for a manned expedition. It is possible that satellite surveys will detect giveaway signs of subsurface life, such as methane diffusing into the atmosphere. However, the best hope for recovering Martian organisms is if they have survived in certain choice locales on or near the surface. For example, a recent cosmic impact might have exposed deep microbe-infested strata; some organisms could remain viable, frozen and inactive, shielded from the Sun's ultraviolet rays by the crater's rim. Another possibility is that ancient halophiles might lie entombed in salt crystals in dried-up lake beds.

NASA Mars expert Chris McKay places his bets on the frozen polar regions, which he thinks may harbor dormant microbes.[4] Although the temperatures there are desperately low, there is at least ice available, unlike in the equatorial regions, which have dried out completely. More clues come from the one place on Earth that resembles the surface of Mars today—Antarctica. In spite of temperatures well below freezing, fierce dry winds, and serious ultraviolet radiation, micro-organisms inhabit the bottom of ice-covered lakes in the McMurdo Dry Valleys. Liquid water can be retained beneath ice even when the average temperature is below freezing through a combination of sunlight, geothermal heat, and the intrusion of meltwater from short episodes when the temperature rises above zero. Martian organisms could have found a final refuge in such a place, and extended their survival time by hundreds of millions of years.

McKay has made a study of an even more remarkable form of Antarctic life. Known by the rather frightening name "crypto-endoliths," these organisms occupy a region within translucent sandstone rocks. They dwell near enough to the surface for light to penetrate, but are protected from ultraviolet radiation and the wind by a thin solid layer. Sunlight absorbed by the rock can create enough humidity from trapped water to keep the organisms going, even at elevations of fifteen hundred meters and temperatures permanently below freezing. Entire communities of bacteria, fungi, lichens, algae, and yeasts live comfortably under these savage conditions. Probably some of these organisms could survive on Mars today using this ingenious strategy, and any indigenous Martian microbes may have evolved similarly.

My own opinion is that the deep subsurface zone of Mars remains by far the most likely location for life today. In fact, for reasons that I shall explain in the next chapter, I believe there is an excellent chance that we will find microbes still living under the Martian permafrost. A few years ago, such a prediction would have been laughed out of court. As long as scientists assumed that life needed sunlight, warmth, and a ready supply of organics to sustain it, Mars seemed a lost cause. However, with the discovery of microbes living in deep, dark, geothermally heated habitats on Earth, the prospects for life on Mars have been transformed.

Meteorites from Mars

In 1911, the little town of Nakhla in Egypt was the scene of one of the most remarkable events in history: a chunk of rock fell from the sky and killed a dog, the only known canine fatality caused by a cosmic object. Improbable though this encounter was already, its truly extraordinary nature was revealed only decades later, when scientists found that the culprit was not a common or garden meteorite, but a piece of the planet Mars. To date, about a dozen Martian meteorites have been identified. Many more are undoubtedly lying around on the ground, their identity unsuspected.

To look at, a Martian meteorite seems little different from any other rock. Indeed, a piece of the Nakhla object was for many years displayed as just another meteorite in the Geology Museum at The University of Adelaide, until its true nature was recognized in the early 1990s. Since then it has been under lock and key. The clue to the Martian origin of these rocks lies not in their appearance as such but in the subtleties of their chemistry. Scientists had long been baffled by a class of meteorites known as Snicks, or SNCs, on account of the unusual amount of volatile material they contained and the strange abundances of their oxygen isotopes. "SNC" is an acronym for the three places of discovery, the "N" standing for Nakhla. The first SNC meteorite to be recovered fell at Chassigny in France in 1815, the second in India, at Shergotty, in 1865.

Most puzzling of all was that the SNC meteorites consist of igneous rock, which is normally associated with volcanoes. This is immediately suspicious. Most meteorites come from the asteroid belt between Mars and Jupiter. Others originate in comets. But comets and asteroids don't have volcanoes; only planets do.

The decisive bit of evidence that there is something odd about the SNC meteorites came in the early 1980s, when they were dated using radioactivity measurements. The ages came out between 180 million and 1.3 billion years. By contrast, ordinary meteorites, which are fragments of primordial material left over from the formation of the solar system, are nearly 4.6 billion years old. A handful of scientists began to suspect that the SNC objects must have come from the surface of a planet—a planet with volcanoes.

Though the discovery of a planetary provenance for the SNCs solved many of the puzzles at a stroke, it raised several others. Most pressing of all was how a large chunk of rock could get off another planet intact and reach Earth. What physical process had the power to eject a rock from a planet without at the same time destroying it? Calculations soon showed that even the most violent volcanic eruption would be unlikely to hurl a rock into space. That left cosmic impacts as the only option. It was certainly conceivable that a planet might be struck by an asteroid or comet with enough force to propel debris into orbit, and that some of this ejecta might eventually reach Earth. However, even in the 1980s, many scientists still found it hard to take such cosmic catastrophes seriously. Moreover, it seemed at the time as if a collision of this magnitude would inevitably pulverize or melt all the rocks in the impact zone. Yet the Snick meteorites have been at most only moderately shocked.

Nevertheless, the weight of evidence steadily accumulated for a planetary origin of the SNCs. The next question was, which planet did they come from? Although Mars was always the prime suspect, confirming it was an exercise in painstaking detective work. Venus is of course a contender, but its thick atmosphere and relatively high surface gravity impede the displacement of material. The Moon— and Earth itself—are possible sources. However, the Moon didn't have active volcanoes as recently as the measured ages of the SNCs. Although Earth did, the meteorites fail the comparison test with both terrestrial and lunar material in a crucial respect: the isotopic ratios of their contents. Not only are the oxygen isotopes all wrong, so are those of xenon, which turn out to be indicative of a planet with a thin atmosphere and a moderately large gravitational field. All this strongly pointed to Mars.

The real clincher came in 1982, in one of those unexpected episodes so often associated with scientific discoveries. NASA scientist Donald Bogard was attempting to date one of the putative Martian meteorites by measuring the abundance of radioactive argon within a pocket of melted glass. Because the results he obtained for his particular sample were clearly absurd, he concluded that the rock had somehow been contaminated. Bogard thought carefully about it, and reasoned that the huge shock wave that blasted the rock off Mars must have forced argon from the atmosphere into the rock. Fortunately, *Viking* had measured the abundances of the argon isotopes in the Martian atmosphere. A comparison immediately showed that Bogard was right. Measurements of the other noble gases, plus nitrogen and carbon dioxide, also agreed with *Viking's* isotope data. The mixture of gases in the meteorite's tiny bubbles matched the Martian atmosphere precisely.[5]

Once it was accepted that the SNCs, and a handful of other meteorites, really did come from Mars, scientists set to work examining them for clues about the physical conditions on the Martian surface. One important discovery was the presence of minerals in the meteorites that have been processed by liquid water, confirming the theory that Mars was once warm and wet. Other information gleaned from the isotope abundances helped build up a picture of changes in the Martian atmosphere. All this work on the Martian meteorites was fascinating and important. Nothing, however, compared to the surprise that lay hidden inside ALH84001.

Traces of life?

NASA has made a startling discovery that points to the possibility that a primitive form of microscopic life may have existed on Mars more than three billion years ago.

NASA ADMINISTRATOR
DANIEL S. GOLDING[6]

The desolate plains of Antarctica are just about the last place you would expect to find meteorite hunters at work. Yet this extensive ice sheet is ideal for yielding astronomical secrets. If you find a stone in the Antarctic ice, there is only one place it can have come from: the sky. Meteorites falling onto the ice soon get buried by snow, but as the ice sheet creeps towards the ocean, carrying the meteorites along with it, it may encounter buried obstructions or rub against mountains. Entombed stones can get thrust to the surface, where they are readily spotted against the white snow.

Roberta Score was a member of the United States Antarctic Search for Meteorites team, and in late 1984 she and her colleagues were given the task of traversing the bleak, windswept glacier near an area known as Allan Hills. About midday on December 27, Score stopped her snowmobile to admire a spectacular ice formation resembling frozen waves. At this point she spotted a meteorite lying exposed on the edge of the ice field. On inspection, it seemed to have a rather unearthly green color. Apart from that, it was just another meteorite as far as Score and her colleagues were concerned, one of over a hundred that the team had collected during their expedition. They were not unduly excited.

As usual, the scientists took care to avoid contaminating their green meteorite—classified ALH after Allan Hills—by placing it in a specially sterilized nylon bag and sealing it with Teflon tape. Nobody touched it with bare hands. It was kept, along with other finds, in frozen conditions during its three-month journey to the Meteorite Curation Laboratory at the Johnson Space Center in Houston, Texas. There it was stored inside a special cabinet in a nitrogen environment to drive off any moisture. It was the first meteorite to be curated among the 1984 batch (hence the 84001 designation) on account of its reportedly unusual color. However, back in the lab it presented only an unexceptional dull grayish hue, and was classified as a common diogenite from the asteroid belt. Thus ALH84001 remained in storage, its significance unrecognized, for four more years.

In the summer of 1988, a geochemist named David Mittlefehldt of the Johnson Space Center was conducting a systematic analysis of diogenites, and he obtained a sample of ALH84001 for analysis. His curiosity had been piqued by the original description that the rock contained certain minerals that are rare in diogenites, such as the weirdly named plagioclase. It was also known to contain carbonates, but Mittlefehldt automatically assumed these must be weathering products from Antarctica.

Mittlefehldt's initial chemical analysis of a bulk sample didn't turn up anything odd. It wasn't until 1990, when he began using an electron microprobe on tiny enclosed grains, that the unique nature of the meteorite slowly became apparent. The probe, which fires a narrow beam of electrons at the surface of the sample and stimulates the emission of X-rays, revealed large amounts of ferric iron, which was quite uncharacteristic of normal meteorites. Mittlefehldt didn't pursue the matter, thinking that his analysis was defective, but in 1993 he wrote a paper on diogenites in which he mentioned the anomalous results from ALH84001. A reviewer of the paper persuaded him to double-check his work. It was only after Mittlefehldt had convinced himself that the chemical analysis was sound that it dawned on him that ALH84001 might not be a diogenite after all, but a Martian meteorite. However, the mineral composition was unlike that of the other known Martian meteorites such as the SNCs. Natural caution got the better of him, and Mittlefehldt didn't mention his conclusion to his colleagues.

The remainder of the story reads like a science-fiction thriller. Mittlefehldt resolved to obtain more samples from ALH84001, but meanwhile turned his attention to another Antarctic meteorite, EETA79002, a straightforward diogenite that he had worked with before. He started using the microprobe in a routine analysis, but was immediately baffled by the discovery, once again, of substantial quantities of ferric iron. As a cross-check, Mittlefehldt studied the composition of iron sulfide in EETA79002, and was astonished to find iron disulfide. "This was totally screwy," he recalled, "because diogenites contain only the iron monosulfide."[7] Totally confused, he went back to basics and examined a thin section of the meteorite under a microscope. It didn't look anything like EETA79002. In fact, it bore an uncanny resemblance to ALH84001. Mittlefehldt checked, and found that he had been given a mislabeled sample; he had been working with pieces of the Allan Hills meteorite all the time! This was proof enough. Iron disulfide is common in Martian meteorites. With the presence of ferric iron, the conclusion could only be that ALH84001 came from Mars.

Once Mittlefehldt had announced, in mid-October 1993, that ALH84001 was another Martian meteorite, it started to receive special treatment. NASA investigator David McKay, also working at the Johnson Space Center, headed up a research group that included Richard Zare of Stanford University. The group began subjecting ALH84001 to a battery of tests. Using fancy chemical and isotopic analyses, the NASA scientists were able to reconstruct a blow-by-blow history of the rock. The first surprise was its age, determined from the radioactive decay of the elements rubidium and samarium. Remember, most Martian meteorites are relatively young, but ALH84001 solidified about four and a half billion years ago, not long after Mars itself had formed.

The investigators paid special attention to the cracks in the meteorite. Evidently something—probably a nearby cosmic impact—had fractured the rock at some stage, partially melting it once again in the process. To determine when, the team carried out careful measurements of potassium-40, a radioactive isotope that decays into argon. Because argon is a gas, it leaks out of molten rock but remains trapped in solid material. The relative amounts of potassium and argon can thus be used to determine the time since the rock cooled down from the shock that created the fissures. The answer came out at about four billion years.

The importance of the cracks in the rock was that deep within them lay tiny grains of carbonate, like limestone. To a geologist, carbonate spells water. The key question was: did this carbonate infiltrate the rock while it rested in the Antarctic ice, or did it come from Mars? The answer was soon forthcoming from the age of the deposits. Although the figure is very rubbery, ranging from 3.6 to 1.4 billion years, the entire range was well before the rock came to Earth.

ALH84001 seems to have had a quiet life until comparatively recently, when a major cosmic impact with Mars kicked it into space. To come up with a date for the ejection, the NASA team studied the effect of cosmic radiation on the meteorite. Material in space is continually bombarded by high-speed particles from the Sun and the galaxy. This radiation produces new isotopes in the material. By measuring their abundances, an estimate can be made of how long the object was exposed to cosmic radiation. For ALH84001, the answer works out at sixteen million years; it spent about this long in space before falling to Earth. To pin down when the rock was blasted off Mars, the scientists needed to determine exactly when the rock arrived in Antarctica. This was done using familiar carbon-isotope dating. Some quantity of the radioactive isotope C^{14} was formed by cosmic radiation when the rock was in space. After it fell to Earth, production of this isotope stopped. By measuring how much had decayed, the time since it fell could be worked out. The answer was about thirteen thousand years. So ALH84001 had lain undisturbed in the ice from roughly 11,000 B.C. until Roberta Score spotted it in 1984.

The NASA team focused their investigations on the distinctive carbonate material in the rock. They knew that these tiny particles would provide important clues about conditions on Mars long ago. Close inspection revealed layered blobs ranging from twenty-five nanometers (one-millionth of a millimeter) up to about a tenth of a millimeter across, coated in iron-rich material that included iron sulfide and a form of iron oxide known as magnetite. All these minerals can be produced separately by different sorts of chemical processes, but their combination in one spot was thought-provoking. What could have made them? After considerable head-scratching, the NASA team began to frame a daring hypothesis. Was it possible that the unusual carbonate grains had been manufactured by living organisms? This was admittedly a wild theory, but had the rock come from Earth, the sort of mineral grains seen would readily be attributed to the activities of microbes.

The investigators badly needed a cross-check: few scientists would be impressed by mineral grains alone as evidence for life. So McKay and his team set about looking for some very different chemicals called polycyclic aromatic hydrocarbons, or PAHs—multiringed molecules known to be produced by decaying life forms. Using a mass spectrometer to search for PAHs, the scientists were rewarded by finding tiny traces of them. Before opening the champagne, however, the investigators first needed to show that these substances hadn't invaded the meteorite while it was in Antarctica. They checked this out by examining the distribution of PAHs within the meteorite, and found that the concentration went up towards the interior—the opposite of what one would expect if the PAHs had infused the rock from the outside. Moreover, other Antarctic meteorites do not contain such quantities of PAHs. This was a tremendous fillip, but it fell short of proving that Martian bugs had been at work in the rock. The problem is that, although PAHs are made by living organisms, they are also made by inorganic processes. Indeed, they have been found in normal meteorites, and even in interstellar space. So their presence in ALH84001 is suggestive but inconclusive. Even if it can be proved that the PAHs come from Mars, they could have been produced by nonbiological processes or delivered there from space.

There was, however, a third reason for the NASA team to suspect that organisms once inhabited the Martian rock, and it was the most dramatic of all. Revealed under a powerful electron microscope were thousands of tiny sausage-shaped blobs clinging to the carbonate grains. These blobs look for all the world like terrestrial bacteria. McKay and his colleagues tentatively concluded that the blobs were nothing less than fossilized Martians—the petrified husks of microbes that lived on the red planet over three and a half billion years ago. If they were right, they would be the first people in history to see the imprint of an alien life form.

Armed with their three pieces of evidence, the NASA team went public in August 1996. The result was an international sensation, with banner headlines around the world and major television coverage spreading the news. President Clinton personally addressed journalists and gave the researchers his blessing. Vice-President Al Gore set about organizing an "implications" seminar at the White House. Religious leaders pronounced soberly on what extraterrestrial life would mean for the faithful. NASA dusted off its Mars-exploration plans and reviewed its budget. The Internet hummed with hastily prepared comment and scientific data. Photographs of ALH84001 were downloaded and used in a hundred impromptu lectures.

I first learned the news in a curious manner. I awoke on August 7 to find a fax addressed to my wife from an acquaintance in England asking whether I was visiting London at that time, for she had just heard me talking on BBC radio about life on Mars. Baffled, I put the fax aside and turned on Australian breakfast television. Sure enough, the big NASA story was being reported. Then it dawned on me what had happened, and I had to smile. For some months I had been giving lectures and interviews around the world about the possibility that microbes or their fossils could reach Earth from Mars (and vice versa) in meteorites. So, when the big story broke, the BBC already possessed an interview with me on this topic, recorded, prophetically, some weeks earlier, before any of us knew about the NASA results. Regrettably the Australian media were less slick. A film crew from the Australian Broadcasting Corporation (ABC) had visited just a month before. I talked to them about the meteorite scenario, and even held up a piece of the Nakhla meteorite for the camera, cracking a joke about getting infected. By a coincidence, this interview was prescheduled for screening on August 8, but the ABC had decided to cut out my bit about microbes in Martian meteorites, because it was either too fanciful or too boring! By the time NASA's discovery was announced, it was too late to put the excised material back in.

An amusing corollary of my precursory interviews was the "white worms" story. In January 1996, I attended a conference in London sponsored by the Ciba Foundation, entitled "Evolution of Hydrothermal Ecosystems on Earth (and Mars?)." At the press briefing, some of us presented arguments for why we thought life on Mars was likely, referring to Martian meteorites and black-smoker ecosystems on Earth with their attendant tube worms. Somehow the message got garbled, and when I appeared on BBC television that evening I was asked to talk about the discovery of white worms on Mars! I did my best to hose the story down, but it never quite went away. To my dismay, when the NASA results were announced, the putative microfossils were widely described as resembling white worms.

Riding the media frenzy, McKay and company kept their cool. Aware that many previous scientific results had been announced amid fanfare only to be retracted, they were careful to stress that the marks in the meteorite fell short of proof that there was once life on Mars, were merely consistent with the hypothesis of a Martian biological origin. More work needed to be done, more information gathered. Only a sample-return mission to Mars would settle the matter definitively. But in their view, life on Mars was the most probable explanation of the facts.

The technical findings eventually appeared in the journal Science,[8] but even before the ink was dry the backlash had set in. Experts leveled several criticisms at the work: contamination by terrestrial PAHs couldn't be completely ruled out; the putative fossils were far too small to be the remnants of bacteria; no "bacteria" had been caught in the act of dividing; the carbonate grains were deposited in conditions too hot to permit life. Some commentators thought the NASA team had been suspiciously lucky. "I've spent my entire career looking for archaean microfossils on Earth," Australian paleogeologist Malcolm Walter remarked to me, "and I've only ever found a handful. Yet these guys find Martian microfossils among a random sample of just twelve rocks!"

The minuscule size of the "fossils" is certainly a powerful objection. At a mere fifty nanometers, the carbonate sausages are a hundred times smaller than most terrestrial bacteria. In fact, they are so small that the question arises whether anything that size could ever have been alive. If they were DNA-based organisms, they could accommodate only a thousand base pairs in their genomes. Even this ignores the existence of any other structures, such as a cell wall, which in terrestrial bacteria is at least twenty-five nanometers thick. Could a Martian microbe perform the alleged mineral-processing feats and other metabolic functions with less than 1 percent of the molecular inventory of a common Earth bacterium? No, say most microbiologists. Yes, say Robert Folk and Leo Lynch of the University of Texas at Austin. Folk and Lynch claim to have discovered, right here on Earth, mineralized microbial forms dubbed nanobacteria that are apparently only a hundred nanometers across.[9] Their claim is supported by the work of a team of Finnish doctors who believe they have isolated living nanobacteria from human blood.[10]

The most serious challenge to the biological interpretation of the features in ALH84001 was made by Ralph Harvey of Case Western Reserve University and Harry McSween of the University of Tennessee. These distinguished geologists examined the meteorite and concluded that the carbonate material was deposited at a temperature of at least 650 degrees Celsius. This would instantly destroy even the hardiest hyperthermophile. But NASA researchers countered the objection with measurements of the oxygen-isotope ratio, arguing that the deposition temperature was no higher than 250 degrees Celsius, and could have been much lower. Unfortunately, their analysis is subject to uncertainty because of the possible loss of the lighter oxygen isotope to space. At the time of writing, this discrepancy has not been resolved, but several other chemical and physical analyses have tipped the balance of evidence against the claim that ALH84001 contains traces of life. NASA has also been accused of exaggerating the evidence for political purposes.

Not all scientists have been so skeptical, however. In fact, a research group at Britain's Open University calmly pointed out that the NASA team were not the first to publish evidence for biological activity in a Martian meteorite. In 1989, Ian Wright, Monica Grady, and Colin Pillinger reported their analysis of another Antarctic Martian meteorite, EETA79001.[11] The British scientists described how they had found organic matter "indistinguishable from terrestrial biogenic components" among carbonate material in the deep interior of EETA79001. And this in a rock less than two hundred million years old. Their findings don't necessarily imply life, but they concluded, knowingly, that "the implications for studies of Mars are obvious."

Killer plague from the red planet!

History may well judge July 20, 1969, to be the most significant date of the twentieth century, the day when human beings first set foot on another world. But when Neil Armstrong, Buzz Aldrin, and Michael Collins returned from the Moon a few days later, they weren't immediately greeted with hugs and kisses. Instead the astronauts were hustled into an odd-looking mobile cabin aboard the USS *Hornet* and left to wave to the world through the window. The purpose of this unceremonious treatment was to quarantine the men, and their cargo of Moon rocks, from the rest of humanity. Although few scientists believed in lunar germs, NASA didn't want to take any chances and inadvertently release a killer plague. In the event, the surface of the Moon turned out to be the most sterile environment yet examined, and the quarantine regulations were quietly dropped for most of the later Apollo missions.

When *Viking* subsequently gave the thumbs-down to life on Mars, the matter of quarantine sank pretty low on NASA's agenda. Today, however, voices of concern are being raised again. If there is life on Mars, and NASA sends a manned expedition there, what if the astronauts bring back virulent Martian bugs? Who knows what the consequences might be? Given the harshness of Mars' environment, Martian microbes might spread like wildfire on our more equable planet. Humans could be totally wiped out by an incurable alien disease, or our crops might be attacked, leading to mass starvation. More insidiously, Martian bugs might gobble up a vital chemical like nitrogen, slowly starving our planet to death. There are certainly many sobering lessons from Earth. When the British settlers released rabbits in Australia, they created ecological havoc. Extraterrestrial germs could prove far more deadly. These fears might soon be put to the test. It doesn't require a manned expedition to Mars to expose us to the risks of alien infection. The return of material to Earth by an unmanned probe—a project well into the advanced planning stage—could present a hazard if surface rocks harbor live organisms or dormant spores.

Although several science-fiction stories have been based on the theme of alien pathogens triggering a deadly pandemic, scientists have generally dismissed such speculation as scare-mongering. They say extraterrestrial microbes are likely to differ so fundamentally from terrestrial organisms that they would pose no real threat. Healthwise, the riskiest micro-organisms are those that most resemble their hosts in their basic biochemistry. In the opinion of Thomas Jukes, a Berkeley biophysicist, "There is no reason to assume that Martian organisms would use the same amino acids or genetic code as does terrestrial life."[12] Martian germs using a different basic system wouldn't even recognize us as living. Jukes points out that nobody is scared of a plague from Antarctica, and justifiably: "Separation *decreases* any such danger by producing divergent evolution." So Jukes infers that Martian bugs should be even more innocuous than Antarctic bugs.

Still, it is better to be safe than sorry. In addressing the contamination risk, NASA long ago determined that they would avoid bringing back any germs from space. A policy document states, "Earth must be protected from the potential hazard posed by extraterrestrial matter carried by spacecraft returning from another planet. . . . Controls on organic and biological contamination carried by spacecraft shall be imposed."[13] More recently, the National Research Council's Space Studies Board instituted a task group chaired by Claude Canizares. Their report, entitled *Mars Sample Return: Issues and Recommendations*, acknowledges that "the risk of potentially harmful effects is not zero," and makes a number of specific safety proposals. For example: "Samples returned from Mars by spacecraft should be treated as potentially hazardous until proven otherwise. No uncontained Martian materials, including spacecraft surfaces that have been exposed to the Martian environment, should be returned to Earth unless sterilized. If sample containment cannot be verified en route to Earth, the sample, and any spacecraft components that may have been exposed to the sample, should either be sterilized in space or not returned to Earth."[14]

It is easier said than done. The actual sterilization process is very problematic. Zapping the samples with toxic chemicals or radiation, or subjecting them to fierce heat, would probably destroy their scientific worth. A task group proposal to coat the external surfaces of the spacecraft with a pyrotechnic substance to be ignited in space seems positively reckless. A more practical suggestion is to expose exterior surfaces to solar ultraviolet radiation. However, the precise method of sterilization has yet to be worked out.

The report also calls for a secure quarantining facility to be established at least two years ahead of time, staffed by experts ranging from microbiologists to earth scientists. Initially, any returned samples are likely to be restricted to a few kilograms of rock, which would be confined to a single containment facility. There will be no passing around bits of Mars rock to interested universities and research labs, as happened with the lunar samples. The material would be screened for signs of bio-activity. Tissue from humans and other organisms would be exposed to the material to test for pathogens. Unfortunately, the costs of providing fully secure facilities to allay all fears could be prohibitive. Some scientists have even pondered the idea of building a containment laboratory in Earth orbit.

John Rummel is a marine biologist at Woods Hole, and a former NASA planetary-protection officer. It was his job to make sure that spacecraft will not inadvertently contaminate Mars with terrestrial microbes and, conversely, that no Martian bugs will be let loose on Earth. So is he worried about the threat of a killer plague coming here in rock samples? He recently told a journalist that, although it was important for NASA to act responsibly, any Martian microbe would find it an uphill struggle infecting well-ensconced organisms like humans, which from the bug's point of view would be totally alien. "I am not sure that anything that might exist on Mars would represent a threat to the Earth," he said, reiterating Jukes' argument. "If you are a human-infecting organism on Mars you would be very lonely."[15] Michael Meyer, the currently serving planetary-protection officer at NASA, agrees. "The odds of bringing something back that will infect humans is practically zero," he is reported as saying, but "it is important to be cautious."[16] Jukes is more sanguine, believing the risks have been overstated: "There is no justification for spending money on quarantining returned Martian samples to protect Earth," he declares.[17]

Although most scientists are dismissive about the danger of Martian germs, the matter is likely to prove of increasing concern to the public. Some groups are already squaring up for legal challenges. "I don't worry about pathogens or things that would infect humans," confesses Rummel. "I think my greatest nightmare is a flock of lawyers stopping a mission because no one has considered the possibility."[18] Jack Farmer, a NASA planetologist and an expert on the possibility of life on Mars, agrees with Rummel: "Planetary protection issues could be a sleeping giant that, once awakened, could dictate the future of Mars exploration."[19]

However much care is taken to minimize the risk of interplanetary contamination, there remains a threat that we can do absolutely nothing about. Martian meteorite ALH84001 comes to us courtesy of Mother Nature. No expensive manned expedition or robot grab was needed to bring the rock to Earth. The small collection of known Martian meteorites represents but a tiny fraction of the millions of Martian rocks that have fallen on Earth, and will continue to fall on Earth in the future. Some estimates suggest that the total amount of Martian material hitting our planet could average as much as a hundred tons per year. If McKay and his colleagues are right, ALH84001 has brought fossilized Martian microbes here. What if another meteorite were to bring live microbes?

Over the past two years, I have been asked repeatedly by the media whether I think the features in ALH84001 really are fossilized Martian bacteria. The question seems a fair one, but it is actually meaningless as asked. Evidence is, as they say, relative. Since the work of McKay and his colleagues falls short of proof, their results can be properly evaluated only in the light of what we already know about the likelihood of life on Mars.[20] If, as the majority of scientists assume, life is the result of an exceedingly improbable accident, then the chances of life's starting independently on Mars (totaling two planets out of two in one star system) are infinitesimal. To such scientists, the marks in the meteorite don't look very impressive. If, on the other hand, one has sound reason to believe that there was life on Mars 3.6 billion years ago, then the evidence presented by NASA is exactly the sort of thing one would expect to find. In spite of the accumulation of negative evidence in recent months, it wouldn't take too much to convince me that ALH84001 contains genuine fossils, because I think there almost certainly *was* life on Mars 3.6 billion years ago. The reason I am so confident in this belief is not that I am sure life emerged from a primordial Martian soup (though it may have), but that the planets are not, and never have been, quarantined from each other.

CHAPTER 9

Panspermia

THINK OF A LOCATION DEEP in outer space, light-years from the nearest star. All around is a black abyss. The temperature hovers just above absolute zero. A yawning void stretches in all directions, populated by a few stray atoms and the occasional fleeting cosmic ray. Through this vast expanse of emptiness there comes, unexpectedly, a solitary grain of matter, too small to be seen with the naked eye. This minute particle drifts unimpeded across the galaxy, heading nowhere in particular. Even through a powerful microscope it would look no more exciting than a speck of dust. On closer inspection, though, this particular speck turns out to be far more than mere dust. It is a bacterial spore.

The spore betrays no overt signs of life. Encased in a thick protective coat, shriveled, dehydrated, and dormant, its very molecules have almost ceased to move, so intense is the cold. It has already been exposed to enough radiation to kill a human being a thousand times over. Yet the spore is not dead, strictly speaking. Nor can it really be considered alive; it does nothing but wait. It may wait a billion years, it may wait forever. But there is an infinitesimal chance that one day the spore will reach a planet with liquid water. Then, suddenly, after a thousand millennia of undisturbed torpor, the spore will return from the dead. Its bacterial soul will begin to stir, genetic memory banks will warm up, metabolism will restart. The bacterium will live life fully once more. And when it does, it will start to multiply—over and over again. A new planet will be seeded with life. That new planet might have been Earth.

This scenario may be entirely fanciful, but it has been taken seriously enough for several recent experiments to be conducted to test it. The idea that organisms can propagate through space has been around for a long time. In 1821, Sales-Gyon de Montlivault proposed that life was triggered on Earth by seeds from the Moon. Shortly after, a German physician named H. E. Richter suggested that meteorites or comets grazing the atmospheres of planets might scoop up floating micro-organisms and convey them to other planets.

At the turn of the twentieth century, the Swedish chemist Svante Arrhenius developed this theory in more detail. He suggested that individual bacterial spores could waft around the galaxy, propelled by the tiny but cumulative pressure of starlight. The nascent Earth, immersed in a rain of dormant but still viable micro-organisms, would have proved a desirable destination for these spacebugs, once the surface was cool enough. Arrhenius called his theory panspermia, meaning "seeds everywhere."[1] It is an idea that has been revisited many times since the original concept was published.

So far in this book I have assumed that, whatever the uncertainty about where and how, Earthlife originated on Earth. But can we be sure? The fact that life established itself on Earth so soon after conditions became favorable has suggested to some that it must have come here from outer space, and that the true genesis of life took place somewhere else in the universe.

Survival in space

Is it credible that unprotected organisms could survive a journey through space? Outer space is hardly a comfortable environment for life. Besides the hard vacuum and low temperatures, there is the radiation: this includes ultraviolet from the Sun, high-speed protons from solar flares, and cosmic rays. Such conditions would soon prove lethal to most known life forms. Yet, in spite of these hazards, not all organisms die quickly in outer space. Bacteria, with their legendary survival capabilities, show remarkable resilience to space conditions.

Scientists from the German Institute for Aerospace Medicine used NASA's Long Duration Exposure Facility to see what happened to spores of *Bacillus subtilis* in space.[2] A series of filters enabled the scientists to test separately the effects of space vacuum, solar and cosmic ultraviolet radiation, and cosmic rays. On retrieval of the samples, up to 2 percent of the bacteria exposed just to the vacuum remained viable. The presence of a sugar or salt layer greatly improved their prospects. Of those exposed to all forms of space radiation, only about one in ten thousand survived, but shielding from solar ultraviolet greatly boosted the survival rate.

Japanese scientists have also proved what a hardy lot microbes are by using a lab simulation experiment, designed to mimic space exposure for 250 years.[3] They sealed up *Bacillus subtilis* spores and other organisms in a vacuum chamber, cooled them to −196 degrees Celsius, and bombarded them with energetic protons from a Van de Graaff generator for twenty-four hours. Half the sample survived this onslaught. The endurance record went to tobacco-mosaic virus, with 85 percent of the sample remaining infectious at the end of the experiment.

Peter Weber and Mayo Greenberg of the University of Leiden in the Netherlands investigated the effects of ultraviolet exposure—the most damaging of all the forms of radiation in space.[4] They cooled spores in a vacuum chamber to −263 degrees Celsius (just ten degrees above absolute zero) to simulate the intense cold of deep space, and shone an intense ultraviolet beam on them from a lamp. The equivalent of twenty-five hundred years' exposure to starlight killed 99.9 percent of the organisms. Nevertheless, a tiny fraction managed to survive. Curiously, the spores seemed to like the cold: their longevity increased markedly at interstellar temperatures.

Such impressive radiation tolerance makes little evolutionary sense unless life has been forced through a radiation bottleneck at some stage in the past. If some microbes have been obliged to adapt to the fierce radiation of outer space, a remnant of this tolerance could survive in terrestrial organisms today. Hoyle and Wickramasinghe cite the case of the bacterium *Micrococcus radiophilus*, which has amazing resistance to radiation, having evolved a special mechanism to repair DNA strands severely damaged by X-rays.[5] This clever little coccus looks very much like the product of an interstellar environment.

Whatever its amazing powers to combat radiation damage, the chances of a live microbe's journeying between star systems would be greatly enhanced if the radiation was at least partially screened. Weber and Greenberg have suggested that microbes might ride to the stars aboard interstellar clouds, which would serve as a type of shield. Such clouds are common throughout the spiral arms of the galaxy; every few tens of millions of years the solar system passes through one. Microbes in the Earth's upper atmosphere, or in impact ejecta, could be swept away with the cloud, perhaps to be transported to another star system. Conversely, any alien microbes resident in the cloud might be transferred to Earth. Typically, the clouds move at about ten kilometers per second and take about a million years to pass from star to star. Although very tenuous by normal standards, they are large enough to block out much of the radiation. Also, a floating microbe might pick up a lot of gunk—ices and organics—in transit, creating an extra protective layer. Weber and Greenberg estimate that the combined shielding effect from cosmic radiation could extend the spores' life expectancy to as much as several million years—time enough to reach another star system.

Trouble for the itinerant spore comes when it approaches a star. Here it gets drenched in ultraviolet. Without a decent coat of absorbent material, death looms. Paul Wesson at the University of

Waterloo in Canada conjectures that panspermia from very old star systems might arrive suitably covered in soot.[6] Stars like the Sun puff out a stream of carbon flecks as they age. A microbe cast adrift in interplanetary space could conceivably get spattered with enough dirt to overcome the ultraviolet risk.

Of course, it isn't necessary for the success of the panspermia process for each and every space-faring microbe to survive interstellar voyages. It demands only a single bacterium to make it alive and find a suitable planet to call home.[7] Life might even be disseminated around the cosmos if the microbes are officially dead on arrival. According to the RNA-world theory, and the experiments of Spiegelman and Eigen I discussed in chapter 5, a primordial soup of chemicals could be triggered into replication simply by adding a suitable RNA template. Just a long fragment of RNA might do to restart the whole process of biogenesis, albeit at a primitive stage. Technically, life would be created anew, but with a vital software ingredient provided from outer space.

Entertaining though these ideas of "naked" panspermia may be, I find it hard to take the theory seriously. The transfer of isolated, exposed organisms between planets is theoretically possible, but the odds are heavily against it. It is most unlikely to be going on systematically all across the galaxy—the radiation risk is just too great. However, there *is* a way for microbes to journey from one planet to another in relative safety, and that is for them to hitch a ride in a meteorite.

Did life come to Earth in a meteorite?

In 1834, the chemist Jöns Berzelius obtained samples of a meteorite that fell near the town of Alais in France. After subjecting them to a careful chemical analysis, he reported on the contents. Most meteorites are stony or metallic in nature, but Berzelius also detected the presence of carbon compounds. Carbon can mean many things; to Berzelius it meant life. Did the carbon in the Alais meteorite imply life beyond Earth? Though Berzelius left the question hanging, later investigators were more forthright. Marcellin Berthelot isolated "coallike" material from the 1864 Orgueil meteorite. Under the microscope, tiny spherical grains were revealed, coated in carbonaceous material. They reminded Berthelot of fossilized bacterial cells.

In the 1880s, the German geologist Otto Hahn came right out and declared that he had discovered a wide range of fossilized life forms buried inside various meteorite samples. These organisms included relatively advanced species such as corals. Hahn's sensational claims were generally discounted. Critics said he just got carried away observing mineral inclusions that superficially resemble living things, like people who see faces in rocks and clouds. Nevertheless, the idea that life came to Earth in a meteorite caught the imagination of many scientists, and claims that meteorites contain signs of life have continued until the present day.

Such claims couldn't be properly tested until scientific techniques improved. By the 1960s, chemical analysis had advanced greatly, prompting Bartholomew Nagy and George Claus in the United States to have another look at the Orgueil meteorite. Using a mass spectrometer, they confirmed the presence of organic material and identified several complex hydrocarbons. But that was just the beginning. Nagy and Claus also reported the discovery of what they termed "organized elements," and appended the sensational conclusion that the organic matter in the meteorite was in all likelihood biological in origin.[8]

As might be expected, Nagy and Claus' claim provoked a storm of criticism. The hydrocarbons were variously attributed to earthly contaminants or mundane chemical processes. Nagy modestly accepted some of this criticism, and called for further studies. There the matter may have rested had it not been for a fortunate event on September 28, 1969, when a meteorite was seen to fall near the town of Murchison in southeastern Australia. The object exploded in midair and showered many fragments over the countryside. Locals began picking up chunks of strange-looking black rock that smelled strongly of methylated spirits. The fall soon came to the attention of John Lovering of the University of Melbourne, who immediately appreciated the nature of the material. The Murchison meteorite belongs to a rare category known as carbonaceous chondrites that are rich in organics; hence the distinctive smell, which persists to this day.

Since their recovery, the Murchison fragments have excited a lot of speculation, and been subjected to a battery of tests that have yielded remarkable results. Among the numerous organic substances found within the meteorite are amino acids that are used by terrestrial life, and some that are not. This raises the obvious question: are these organic substances the decayed remnants of extraterrestrial organisms, or did they form from some simple chemical process? A factor that has a bearing on this question is the discovery that some of the amino acids in the Murchison meteorite have an excess of left-handed over right-handed helicity. As I explained in chapter 3, one of the distinctive features of life on Earth is that it manufactures right-handed DNA. Similarly, it uses amino acids of the left-handed variety only, so the helicity bias in the meteorite could suggest a biological origin. On the other hand, there are known to be physical processes (such as illumination with polarized light) that could also enhance the synthesis of left-handed amino acids.[9]

The Murchison meteorite proves one thing at least. There are objects in space loaded with just the sort of organic compounds needed for life to get started. It doesn't require a primordial soup on Earth to synthesize the building blocks of life. These substances can fall from the sky, ready-made.

Did Earthlife come from Mars?

Though I have spent a lifetime in scientific research, I cannot remember having had more than a dozen truly original thoughts. The good ideas I did dream up usually entered my consciousness only gradually, and congealed bit by bit while I was immersed in my work. Sudden revelations and blinding flashes of insight are, I believe, actually rather rare in science. One memorable occasion when I did put two and two together on the spur of the moment occurred in July 1992, when I was attending a lecture by Lloyd Hamilton at a meeting of the Australian and New Zealand Association for the Advancement of Science in Brisbane. Hamilton was talking about the subterranean biosphere, discussing his work on organisms that live in the Earth's rocks. Partway through his lecture, it occurred to me that, if apparently solid rocks can have microbes living inside them, and if rocks can travel from Mars to Earth (or vice versa) in the debris of cosmic impacts, then microbes could tag along, and so cross-contaminate the planets. With a rock as a protective shield, the journey would be far less hazardous than it would be in Arrhenius' theory. In particular, organisms might readily be conveyed between Earth and Mars that way. I raised the matter in the question time at the end of the lecture, but my proposal seemed a bit wild and the discussion didn't proceed very far. Nevertheless, I continued to develop the idea over the coming months, and lectured on it at the University of Milan in November 1993. Again, the response was muted. Undaunted, I included the conjecture in my little book *Are We Alone?*, which was published the following year. Sometime after this, I discovered that Jay Melosh of the Lunar and Planetary Laboratory of the University of Arizona had independently come to similar conclusions.[10]

Alas, there is nothing new under the Sun. Melosh and I were by no means the first to spot the possibility that microbes might travel between planets inside ejected rocks. As early as 1871, no less a scientist than Lord Kelvin pointed out that the collision of an astronomical body with a planet might displace much debris, and thus "many great and small fragments carrying seed and living plants and animals would undoubtedly be scattered through space." In an address to the British Association in Edinburgh, Kelvin surmised that some of these fragments would eventually reach other planets and infect them with life:

Because we all confidently believe that there are at present, and have been from time immemorial, many worlds of life besides our own, we must regard it as probable in the highest degree that there are countless seed-bearing meteoric stones moving about through space. If at the present instant no life existed upon this earth, one such stone falling upon it might . . . lead to its becoming covered with vegetation.[11]

If life can indeed hop from one planet to another, we cannot be sure that Earthlife began on Earth. It could, for example, have come from Mars.[12] We know that microbes live deep within rocks on Earth. If there was life on Mars too, it is likely to have begun in the form of chemotrophs living underground. So rock fragments ejected by a cosmic impact with Mars might well contain micro-organisms inside them. Cocooned within a rock, live Martian microbes could be successfully transferred to Earth.

At first blush the theory seems to suffer from a major snag. Wouldn't a blow big enough to blast a rock into space instantly reduce any microbial inhabitants to pulp? Curiously, the answer is no. The microbes are saved by their very smallness. Let's put some numbers in. An impact with the power to project a rock from Mars at escape velocity (five kilometers per second) would subject the microbes to ten thousand g acceleration. Such a huge impulse would certainly crush most organisms. However, the tiny dimensions and low mass of micro-organisms means that they could probably withstand such colossal g forces and leave Mars relatively unscathed.

There is a more serious hazard from the ejection process, however. A large cosmic impact generates a huge shock wave that compresses the surrounding rock. Like all material, rock heats up as it is squeezed, and even modest compression would raise the temperature to lethal values. Until recently, geologists assumed that a cataclysm capable of launching a rock into orbit would also melt it. Laboratory experiments indicated that the ejected material would suffer massive compression—equivalent to an applied pressure of at least one and a half megabars. But the discovery of Martian meteorites proved this view to be mistaken: although some of them show signs of moderate shock-heating, others have clearly escaped from Mars pretty well undisturbed.

Jay Melosh has tackled the problem of how a rock can be impelled into space without being destroyed in the process. He worked out a detailed mathematical model of a cosmic impact, and sketched out the following picture of events. First the incoming asteroid or comet punches a hole in the ground. The energy released is so great that most of the impactor itself is vaporized. Directly beneath ground zero, the explosive release of energy squashes the rock, vaporizing or melting much of it as a result. The compression wave then propagates laterally, spreading out through the surrounding terrain and deep into the ground. The elastic energy stored in the subsurface rocks as a result of this compression is then liberated again as the material rebounds, and this delivers an enormous vertical force to the overlying layers. Unlike the lower strata, the surface rock cannot be compressed, because it is free to move upwards; the only restraining force is atmospheric pressure, which is negligible. Therefore, the surface material shoots skywards without getting crushed, and if the force is great enough it will fly right off into space. The impact crater is not so much a dent in the ground as a cavity excavated by the sudden upthrust, and will be many times bigger than the size of the impactor. Much of the material near the edge of the crater simply gets flung aloft rather than squashed down by the blow. An added bonus is that, because the incoming object evacuates a huge tunnel in the atmosphere, the ejected rocks are spared frictional heating as they take off. Melosh thinks that the peripheral rock first ascends as a continuous plate of material, pushed from below, and then shatters into fragments. His calculations predict that the size of the pieces depends on the overall scale of the blast. On the whole, a bigger bang ejects larger fragments. A major impact would launch several million rocks of about ten meters in diameter. Some of them would get very hot, but a good fraction would stay below a hundred degrees Celsius.

Perhaps surprisingly, then, a microbe inside a Martian rock

could get into space without being killed by either blast or heat. However, its problems would only just have started. Once it was in space, its fortunes would turn on the exact trajectory of its rocky spaceship. Much of the ejected debris would go into orbit around the Sun. Because a rock in interplanetary space is subject to the gravitational forces not only of the Sun but of the planets too, its motion can be very complicated, even chaotic. Each time a Martian rock nears Mars on its journey round the Sun, it will receive a tug from the planet's gravity. After many such tugs, the rock may be propelled into an Earth-crossing orbit, or hurled towards the outer region of the solar system, where it will become subject to gravitational perturbations from larger planets. The rock may spend a long time in this cosmic pinball machine before its ultimate fate is decided.

What are the chances that a Martian rock will reach Earth? A recent computer calculation predicted that 7.5 percent of rocks ejected from Mars will be swept up by the Earth eventually, a similar proportion going to Venus.[13] Most of the debris (38 percent) ends up in the Sun, 9 percent recollides with Mars, and much of the remainder goes off towards Jupiter and gets flung right out of the solar system. The sojourn times in space are surprisingly short. About a third of the rocks delivered to Earth arrive in the first ten million years. These results compare well with the times in space of the known Martian meteorites, which can be measured from their cosmic-ray exposure. Measurements range from fifteen million years for ALH84001 down to a mere seven hundred thousand years for EETA79001. Some Martian meteorites will inevitably get here very fast if they leave Mars at a favorable speed and angle. In the simulation, transfers as rapid as sixteen thousand years were observed. Simply on statistical grounds, a few rocks will make the journey in less than a century.

Whether a microbe in a Martian rock will reach Earth in a viable condition depends on how long it can remain alive in space. Of course, we know nothing about the putative Martian microbes, but if the endurance feats of terrestrial bacteria are anything to go by, they might hold out for a very long time. Archaeologists occasionally unearth a tomb containing bacterial spores dating from the time of the tomb's construction. There is a report of live E. *cloacae* found in a mastodon skeleton eleven thousand years old. Under deep-freeze conditions, much longer survival times are possible. Chris McKay has discovered three-million-year-old micro-organisms preserved in the Siberian permafrost. Amazing claims have been made about bacteria preserved in salt beds for hundreds of millions of years. Forty-million-year-old bacteria have also been extracted and cloned, in true *Jurassic Park* style, from bacteria inside a bee trapped in amber.[14]

The British microbiologist John Postgate, who has made a study of bacterial mortality, has questioned whether bacteria need die at all.[15] When starved of nutrients, they slowly become senescent; their metabolism grinds to a halt, they shrink dramatically in size, and they stop reproducing. But they do not necessarily die in the normal sense of the word; they may simply slip into a state of suspended animation. If conditions one day improve, they can be resuscitated, like Sleeping Beauties. There is no known internal clock that determines a "point of no return." In fact, it is quite mysterious precisely what decides whether or not a given spore can be revived from this nearest of near-death experiences. If nothing is going on inside, what enigmatic line marked "death" must be crossed before revival is impossible?

Bacterial immortality assumes that nothing damages the microbe's vitals irreparably. One obvious cause of damage is radiation. Although bacteria have repair mechanisms that can combat radiation damage, these don't work in the dormant state. If the DNA of a dormant microbe gets fractured, it stays fractured. For a journey through space, radiation is undoubtedly the greatest danger. However, most of the radiation would be unable to penetrate a rock. Ultraviolet is absorbed by a thin layer, and all but the highest-energy cosmic rays will be stopped by a meter of solid material. The rock itself will have some residual radioactivity, but, as we have seen, bacteria are astonishingly resistant to radiation damage. Dehydration—inevitable under the vacuum conditions of space—seems to afford added protection. It would take millions of years for an organism inside a large rock to be subjected to a lethal dose of radiation. This is easily long enough for it to travel from Mars to Earth.

Another factor to consider is the cold. In interplanetary space, the temperature is low but not extreme. The rock would still be warmed by the Sun, and the internal temperature would be likely to bottom out at around minus fifty degrees Celsius. This is perfectly acceptable; indeed, bacteria are routinely stored in refrigerators at much lower temperatures. If anything, the cold of outer space would prove an advantage in preserving the microbes.

Though the journey through space is far less hazardous than it may have seemed at first sight, the tribulations of a Martian microbe are not over when it reaches Earth. It still has to avoid being killed when the rock plunges through the air at many kilometers per second. Most small meteorites burn up completely on entry into the atmosphere. However, for a one-to-ten-meter rock arriving on a shallow trajectory, it is a different story. The rock would be slowed by air friction and possibly explode as a result of the buffeting, showering fragments into the upper atmosphere. The pieces would then fall to the surface at terminal velocity, which is relatively gentle. Disintegration would spill out some of the microbes in midair; others would reach the ground or the ocean still protected within fragments of rock. Many meteorites have been known to fall this way. Because rocks are good heat insulators, the interior of the meteorite will remain cold even when the outer surface is melted by friction. All in all, these circumstances are more or less ideal for the successful dissemination of any resident organisms.

Once it is safely down, the prospects for a Martian microbe would depend on the conditions it encountered. Three or four billion years ago, when Mars resembled Earth, a Martian organism might well have found our planet very much like home, especially if it fell in the sea. Carried along by ocean currents, it might eventually have reached a deep ocean volcanic vent similar to its original habitat on Mars.

Some people think that the chain of favorable circumstances needed to bring a Martian microbe safely to Earth stretches credulity too much. Certainly at each step of the way only a fraction, and perhaps only a very tiny fraction, of ejected microbes would survive. But the journey doesn't need to be comfortable, only survivable. It needs just *one* Martian chemotroph among the trillions that may have been ejected to make the journey alive, and the way would lie open for wholesale colonization of Earth. A large impact event of the sort that has left both Earth and Mars littered with craters would blast billions of tons of material into space. Millions of rocks a few meters across would be scattered around the solar system, many of them candidates for transporting life. Bigger impacts would produce even more ejected debris. At the end of the heavy-bombardment period, the numbers would be higher still. It is hard to avoid the conclusion that, if there was life on Mars between three and a half and four billion years ago, then live Martians will inevitably have taken up residence on Earth. This possibility has also occurred to the U.S. National Research Council's Space Studies Board. In their *Mars Sample Return* report, they state: "Survival of microorganisms in a meteorite, where largely protected from radiation, appears plausible. If microorganisms could be shown to survive conditions of ejection and subsequent impact, there would be little reason to doubt that natural interplanetary transfer of biota is possible. . . . Such exchanges would have been particularly common early in the history of the solar system when impact rates were much higher."[16] Whether any Martian microbes successfully colonized Earth following their arrival is, of course, another matter to which I shall come shortly.

Is there any reason to favor Mars over Earth as the cradle of life? One factor suggests the answer is yes. The same cosmic bombardment that provides a mechanism for transporting organisms between the planets also threatens their survival at home. As I explained in chapter 8, a really big impact will effectively sterilize an entire planet. In this respect, Mars may have been safer than Earth. Its smaller size would make it less of a target for asteroids and comets. The lower gravity would result in slower impacts that did less damage, allowing useful organic material to accumulate. In particular, Mars was spared the really huge impact that created Earth's Moon. The heat of formation of Mars was also less than that of Earth, so all in all it seems clear that Mars would have cooled off more quickly, making the planet habitable perhaps as early as four and a half billion years ago. Mars' cooler crust would also mean that the comfort zone for a subsurface microbe would go a lot deeper, providing a better refuge against the heat pulses from impacts.

The deep subsurface zone may not be the only available refuge from cosmic bombardment. Another is outer space. The impact event that sterilizes the planet will also splash a huge quantity of material into the relative safety of planetary orbit. If microbes can survive in space inside these displaced rocks, then some may eventually return to reseed the planet, after it has recovered from the effects of the cataclysm. Because Mars has a lower escape velocity than Earth, material can be ejected with less violence: microbes are more likely to survive. Towards the end of the period of heavy bombardment, Mars must have been enveloped in a swarm of ejected debris, which might have harbored large numbers of exiled organisms.

Not only was Mars a better place for life to start, it could also have proved a more favorable location for life to evolve. Biologists suspect that life on Earth really took off only after oxygen became available in the atmosphere, about two billion years ago. At that stage, there was a rapid diversification of species. On Mars, it is likely that oxygen built up much faster, perhaps in as little as ten million years. It could be that, even before the end of the heavy bombardment, life on Mars had evolved to a level not attained on Earth for another billion years.[17]

If life arose independently on Earth and Mars, a Martian microbe reaching Earth might arrive to find organisms already well ensconced. The newly arrived transportees would then be pitted in competition with their terrestrial counterparts. The Martians might get gobbled up by Earthly bacteria as soon as they arrived. This would be a cruel twist of fate for such intrepid travelers. To survive being hurled into orbit by a cosmic impact, to endure millions of years in outer space, to avoid being burned up on re-entry, and to be lucky enough to drop near a suitable habitat, only to end up as a meal for a foraging rival, would be the supreme irony.

Other scenarios are possible. The Martian microbes might eat the terrestrials rather than the other way around. Then again, Martian and terrestrial microbes might occupy different niches and peacefully coexist. If they were based on radically different biochemistry, they could happily ignore each other. Or they might be very similar and get on well enough to engage in symbiosis (e.g., Martian mitochondria invading terrestrial bacteria). It could even be that you and I have some Martian genes in our bodies! Or perhaps the interlopers found the going too tough on Earth, failed to adapt in time, and died out after a brave attempt at colonization, like the early settlers of some Australian outback towns.

It is conceivable that Martian microbes still exist on Earth as an independent life form. Scientists are only beginning to discover the huge number of micro-organisms that live all around us. So far, all those that have been discovered are related to terrestrial life, but one day a truly alien micro-organism may be unearthed, perhaps in a weird or inaccessible place—deep under the ground, or in the upper atmosphere, or even under the Antarctic ice cap. If alien microbes use a different biochemistry, scientists may have failed to spot them. They might even be lying dormant all around us, in sporelike form, unable to achieve revival for lack of some crucial ingredient.

The foregoing suggestions are, of course, pure speculation. All we can say for sure is that, if there is, or was, Martian microbial life, then a viable Martian microbe will almost certainly have made it to Earth at some stage in the last four billion years. Whether life actually originated on Mars and spread to Earth is more conjectural. If it did, we are led to a curious thought: you and I, and all the living things we see around us, would be descended from Martians.

Did Earthlife go to Mars?

If microbes can travel from Mars to Earth in rocks, they can go the other way too. Although Earth has a stronger gravitational pull, there have certainly been cosmic impacts powerful enough to fling terrestrial material into space. In this case we know that some of the ejected rocks must have contained micro-organisms. If there really was life on Earth at the end of the period of heavy bombardment, as the fossil evidence suggests, then copious quantities of life-bearing material will have been put into space by the many very large impacts that were still occurring 3.8 billion years ago. Some of this material will definitely have reached Mars, at a time when conditions there resembled those on Earth. It is therefore inevitable that life from Earth has reached Mars at some stage during its history. It also appears extremely likely that between 3.5 and 3.8 billion years ago conditions on Mars would have been suitable for transported terrestrial organisms to flourish. That is why I am certain that there was life on Mars in the past, and may well be life there today.

When the story broke about the NASA meteorite, commentators and scientists alike jumped to the conclusion that life must have happened twice in the solar system. The features in ALH84001 were almost universally taken as evidence for the independent origin of life on the planet Mars. The profound philosophical conclusions that Clinton and others rushed to draw—of a universe teeming with life, of bio-friendly laws operating throughout the cosmos—depended crucially on that tacit assumption. Remarkably few people spotted the basic flaw in the logic: if a fossil Martian microbe can come to Earth in a rock, then a live microbe can go from Earth to Mars in a rock. The very source of the evidence for life on Mars itself undermined the independent-origin theory.

If life did reach Mars from Earth, that would certainly be exciting and scientifically important, but it would have zero philosophical significance, for it would tell us nothing new about the uniqueness or otherwise of the phenomenon of life. It would simply show that the Earth's biosphere extends out into space as well as under the ground. The putative microfossils in ALH84001 would then be the descendants of what were originally terrestrial organisms anyway, returning home.

The likelihood of planetary cross-contamination, especially in the remote past, is a key factor in assessing the evidence for life on Mars. If Mars was inoculated with Earthlife 3.6 to 3.8 billion years ago, it would be no real surprise to find Martian rocks containing signs of life at work 3.6 billion years ago. As I stated in the last chapter, the features in ALH84001 are exactly what one would then expect to find. On the other hand, if the contamination theory is wrong, the rules of the game change dramatically. We are then being invited to believe that life started independently on Mars—a *huge* assumption, which requires considerable justification (see chapter 10). In that case, the evidence of ALH84001 is far, far less persuasive.

How can the contamination theory be checked? If scientists could get hold of a live Martian organism, and it turned out to be based on right-handed DNA and left-handed amino acids, if it had the same genetic code as Earthlife, and if its metabolism were also similar, this would point strongly to a common origin with life on Earth. On the other hand, if it had oppositely handed molecules or a different genetic code, or was based on a completely different form of biochemistry altogether, then an independent origin would be implied. Settling the matter would be harder if fossils are all that remains. Remnants of familiar organic molecules with opposite handedness would still give the game away, but simply comparing microbial shapes is no help. Alien microbes might look like terrestrial microbes but have a completely different biochemistry.

Suppose, as I am claiming, that life-harboring material has been regularly exchanged between Earth and Mars; then these planets cannot be considered quarantined. Cross-contamination might have been going on since life first started. If so, it seems pointless to go to the expense of sterilizing our spacecraft. Conversely, we can no longer assume that the risk of infection from Martian micro-organisms is negligible. If Martian and terrestrial life forms are descended from a common ancestor, Martian microbes will have the same basic biochemistry as we do. As Carl Sagan wrote: "If putative Martian organisms were originally transferred to Mars by collisions with Earth, they may be enough like us that they could be pathogenic."[18]

If Earth and Mars have exchanged organisms, it considerably complicates the question of where life ultimately began. Given our present state of ignorance, it is an open bet which of the following scenarios might be the truth:

1. Life began once, on Mars, and came to Earth in Martian meteorites. It may or may not still exist on the originating planet.
2. Life began once, on Earth, and was propagated to Mars, where it possibly established itself.
3. Life originated on both Earth and Mars independently. Cross-colonization (or even cross-fertilization) may subsequently have occurred.
4. Life originated on both Earth and Mars, but in spite of the exchange of rocks and dust, no transfer of viable organisms has occurred.
5. Life originated on neither Earth nor Mars, but somewhere else entirely, such as a comet, Jupiter's moon Europa, Venus, or a body outside the solar system altogether. It came to Earth, and perhaps Mars too, via some sort of panspermia mechanism.
6. Life originated on Earth alone and has not (yet) successfully colonized another planet. Mars is, and always was, lifeless.

Notice that, apart from the final one, all of the above scenarios predict that there must have been life on Mars once, and may very well be life there today. From what we know about the incredible staying power of microbes, I think scenario 6 is very improbable. At some stage, rocks carrying viable organisms must have made the trip from Earth to Mars. Either way, whether life began more than once or simply propagated between planets, it seems to me inevitable that Mars has played host to microbial populations, and perhaps to more advanced organisms too, at an early stage in its history. This makes the search for life on Mars an urgent priority.

If the idea of rocky panspermia is accepted, then Mars is not the only planet of interest. It is conceivable that Earthlife has traveled elsewhere in the solar system. How about the Moon? Today the lunar surface is extremely unpleasant, but, like Mars, it once had a thick atmosphere, volcanoes, and water. These disappeared faster than on Mars, but there may have been a short window of opportunity for life. If that window coincided with life's presence on Earth, the chances of an exchange of organisms to the Moon must be considered very high. Given the Moon's proximity to Earth, a lot of terrestrial impact ejecta ends up on the Moon, and the travel time is very short. Could there be life on the Moon today, under the surface? Recent claims that ice may exist at the lunar poles in craters shielded from the Sun raises the intriguing—though extremely speculative—prospect that live lunar microbes may yet be found.

Venus and Mercury look like lost causes, since both are far too hot. Possibly Venus was once cooler, and might have provided a home for displaced terrestrial organisms for a time. Several moons in the outer solar system might just possibly provide suitable abodes for life, although the chances of a successful transfer of life from Earth are remote. Thomas Gold has conjectured that at least ten planets or moons might support life beneath the surface, and that subsurface life may be very common in the universe. He believes that on Earth we may have "just one strange branch of life," because the unusual conditions here have made surface life possible.[19]

Just as rocks blasted into Mars orbit could have provided a refuge from cosmic bombardment, so terrestrial organisms might have been displaced into space, only to return millions of years later to recolonize Earth. This possibility puts a new slant on the impact-frustration theory of life, which I discussed in chapter 6. Impacts that would totally sterilize Earth might yet leave organisms alive amid debris in Earth orbit. This enables us to push back the date for life on Earth well into the period of heavy bombardment, perhaps as far back as 4.2 billion years, and helps alleviate the paradox that life apparently existed on Earth through such a violent epoch. Of course, if ancient microbes can return after millions of years off the planet, there are intriguing implications for evolutionary history. It is not impossible that even today a ten-million-year-old bacterium, extinct on Earth, will return in a meteorite and re-establish itself.

What about the possibility of life traveling between the stars inside rocks? Unfortunately, the statistics weigh heavily against this idea. Whereas material blasted off Earth stands a reasonable chance of hitting Mars, the likelihood is negligible that a rock ejected from the solar system will ever encounter another planet. The distances between the stars are so great, and planets are such small targets, that even with billions of rocks being scattered into the galaxy there is little prospect that one of them will drop onto a suitable planet in another star system. For the same reason, it is almost certain that no rock carrying life from another star system has ever hit the Earth. So, whereas the planets within our solar system might well have contaminated each other with life-bearing rocks, it is extremely unlikely that life could spread across the galaxy this way.

However, rocks are not the only vehicles in which microbes could hitch a ride. Comets might serve the same purpose. Although little is known about the interiors of comets, they could provide an even better refuge for microbes than rocks. This would certainly be the case in the period immediately after the formation of the comets, when chemical and radioactive heating may have raised the temperature enough to support liquid water.

Chris McKay envisions a scenario involving cometary panspermia that goes as follows.[20] An interstellar cloud arrives in the vicinity of the solar system. Comets, perhaps perturbed by the cloud's gravitational field, bombard Earth and eject into the cloud debris containing microbial spores. The displaced rocks stay with the cloud until, maybe millions of years later, the cloud itself starts to spawn stars. When this happens, the rocks, together with some still-viable organisms, mingle with the cometary material near the edge of some new star system's nebula. As the comets form, they provide warm, wet interiors in which the long-suffering microbes might at last be comfortably incubated. The improved environment revives the spores and triggers explosive multiplication. Sometime later, one such comet plunges towards the star, complete with its now extensive microbial colony. The star evaporates the comet's material and releases the microbes. They spew out along with the cometary dust in their countless trillions, forming a vast living cloud. Although the organisms are now horribly exposed and vulnerable, it is not long before some of them are swept up by a planet passing through the comet's tail. Being so tiny, the microbes survive re-entry, and drift slowly downwards to the relative safety of the planet's surface. In this way, life might colonize planets in other star systems. Equally, it might have come to Earth this way from a planet beyond the solar system.

For the past twenty years, Fred Hoyle and Chandra Wickramasinghe, in the face of much skepticism, have been pushing the theory that comets contain living organisms. They offer support for their idea through an analysis of medical records, and claim that the passage of comets is correlated with the outbreak of diseases.[21] They suggest that various pandemics, such as the great Justinian plague of A.D. 540, in which possibly a hundred million people died, are actually of extraterrestrial origin. Hoyle and Wickramasinghe do not suppose that life in space is restricted to comets. They back Arrhenius' original proposal that individual microbes can float unprotected around the galaxy. Pointing to the fact that many interstellar grains are about the size of bacteria, they argue that substantial quantities of material in interstellar space are actually of biological origin. As evidence for this audacious theory, they cite the fact that the infrared spectrum of dry E. coli looks uncannily close to that of interstellar dust.

Not surprisingly, some scientists have seized on the panspermia theory in an attempt to evade the problems of biogenesis. If life can propagate between star systems, then only one planet is needed to spawn life, somewhere in the vastness of the cosmos, to account for the existence of life on Earth. I do not share this enthusiasm for evasion. It seems to me that shunting the problem off into outer space does nothing to address the central problem of biogenesis—the problem that has plagued researchers in this discipline for decades—which is that life seems just too good to be true.

CHAPTER 10

A Bio-Friendly Universe?

*The more I examine the universe and study the
details of its architecture, the more evidence I find
that the universe in some sense must have known we
were coming.*

FREEMAN DYSON[1]

W HEN THE CRIPPLED GALILEO SPACECRAFT painstakingly
beamed back pictures of Europa from its backup antenna in
April 1997, NASA scientists were jubilant. The word on everybody's lips was—"Life!" The excitement focused on the discovery of
the first known extraterrestrial ocean. Europa, scientists already
knew, is covered in ice. What *Galileo* revealed were icebergs. Icebergs spell liquid water, or at least slush. The entire frozen crust of
this frigid Jovian moon seems to be slithering around on a layer of
fluid.

Almost to a man (and woman), commentators intoned that water plus organics means life—or at least a good chance of it. The ra-
tionale was summed up by NASA mission scientist Richard Terrile.
"Put those ingredients together on Earth and you get life within a
billion years," he told the press.[2] Ergo, it will happen on Europa too.
Just like that, as the British magician Tommy Cooper used to say.
Unfortunately, the slender thread of logic that links water and life is
scarcely more than the observation that life without water seems
impossible. Equating water with life conceals a gigantic leap of faith.

It may be that life does indeed reside beneath Europa's icy skin, either for the relatively trivial reason that it traveled there from Earth in a meteorite, or for the much more profound reason that life is inevitable given the right conditions. According to the deterministic school of biology, which seems to dictate the prevailing view at NASA and is shared by most media commentators, life will automatically form in any Earth-like environment. Take a measure of water, add amino acids and a few other substances, simmer for a few million years, and—hey presto!—it lives. This popular theme is sharply criticized by the opposing school, which stresses the awesome molecular complexity of even the simplest living thing. To proponents of the latter position, the sheer intricacy of life bespeaks a freakish concatenation of events, unique in the cosmos. No amount of water, they say, even if laced with fancy chemicals, will come alive on cue. Earthlife must therefore be a fluke of astronomical improbability.

In claiming that water means life, NASA scientists are not merely being upbeat about their project. They are making—tacitly—a *huge* and profound assumption about the nature of nature. They are saying, in effect, that the laws of the universe are cunningly contrived to coax life into being against the raw odds; that the mathematical principles of physics, in their elegant simplicity, somehow know in advance about life and its vast complexity. If life follows from soup with causal dependability, the laws of nature encode a hidden subtext, a cosmic imperative, which tells them: "Make life!" And, through life, its by-products: mind, knowledge, understanding. It means that the laws of the universe have engineered their own comprehension. This is a breathtaking vision of nature, magnificent and uplifting in its majestic sweep. I hope it is correct. It would be wonderful if it were correct. But if it is, it represents a shift in the scientific world-view as profound as that initiated by Copernicus and Darwin put together. It should not be glossed over with glib statements that water plus organics equals life, obviously, for it is far from obvious.

If biological determinism is indeed confirmed by the discovery of alternative life beyond Earth, it will dramatically confound the orthodox paradigm, steeped as it is in Darwinian contingency. Orthodoxy insists that nothing in life is preordained, that biological evolution is a long series of meaningless, directionless accidents: there are no final causes. But if life is somehow inevitable, accidents of fate notwithstanding, a particular end is certain to be achieved; it is built into the laws. And "end" sounds suspiciously like "goal" or "purpose"—taboo words in science for the last century, redolent as they are of a bygone religious age.

The ramifications of finding life elsewhere in the cosmos are therefore profound in the extreme. They transcend mere science, and have an impact on such philosophical issues as whether there is a meaning to physical existence, or whether life, the universe, and everything are ultimately pointless and absurd. That is the momentous import of the search for life on Mars and beyond. That is why we should pursue that search as a matter of the highest priority. And that is why the panspermia theory is so crucial. To prove a bio-friendly universe, we have to know for sure that life has happened more than once, which means ruling out planetary cross-contamination as the explanation for any extraterrestrial organisms that may be discovered. Finding Earthlife on Mars would tell us nothing new about the origin of life. But if contamination can be discounted, just a single Martian microbe would transform forever our picture of the cosmos.

The search for life in the universe is thus a search for ourselves—who we are and what our place is in the grand scheme of things. So, what does the scientific evidence suggest? Are we just insignificant freaks, or the expected products of an ingeniously bio-friendly universe?

Did life ever begin?

The entire discussion about the origin of life proceeds from the assumption that life actually had an origin. Is it conceivable that life has always existed? Clearly Earthlife has not always existed, because the Earth itself has not always existed. But life may have been around before the Earth formed, and have come here by some panspermia process. If organisms are able to propagate from star to star across the universe, then the question of whether life had a beginning reduces to the question of whether the universe had a beginning.

In the nineteenth century, most scientists supposed that the universe was eternal. It was then possible to believe that life is coextensive with the universe in both space and time. This was the position championed by Svante Arrhenius and Lord Kelvin. Today, most scientists believe that the universe has not always existed, but began in a big bang. There is good observational evidence to support that theory. However, there is no known fundamental reason why the universe cannot have always existed. A model of a universe with neither beginning nor end, known as the steady-state theory, was popular in the 1950s. Its principal proponent was Fred Hoyle. Both the big-bang and the steady-state theories assume that the universe is expanding. In the big-bang theory, all the cosmic material comes into being more or less in one go at the outset. As the universe expands and the galaxies fly apart, so the average density of matter declines. By contrast, in the steady-state theory, the average density remains constant. Matter is continually created, forming into new galaxies that occupy the widening spaces between the old ones. On a large scale, the universe stays much the same from epoch to epoch, like an ever-replenished well.

Because a steady-state universe has infinite age, we can imagine that life might also have existed forever. Then neither cosmos nor life would have had an origin. As long as there is a way for organisms to get from old galaxies to new ones, life need never have formed *de novo* from inert chemicals. The problem of biogenesis is therefore completely sidestepped. It isn't necessary to adhere to the steady-state cosmology as such to avoid an origin of life. Provided that the universe is infinitely old and has some sort of replenishment process, and as long as microbes can find a way to travel safely from one place to another, then life may always have been a property of the universe. In fact, that is precisely what Hoyle and Wickramasinghe propose.[3]

The theory of eternal life does have a rather curious corollary. If life extends throughout space and time, and if, as would be the case in a steady-state universe, there is an infinite number of planets, then there will be an infinite number of biosystems. If a fraction of those biosystems develop intelligence and technology, there will be an infinite number of technological communities in the universe. Because there is no limit on how long ago such technological communities may have arisen, some of them will be arbitrarily ancient and arbitrarily advanced. If microbial life can spread across the cosmos, so can advanced intelligent life. We are thus drawn inexorably to the outlandish conclusion that the universe must have been "taken over" by intelligent life. It requires but one expansionist technological community of unlimited age for intelligence to gain control of the cosmos. Indeed, given the infinite amount of time available for this process to happen, nature and technology will by now have effectively become as one. So intelligence will also be coextensive with the universe. Mind would be just as much a permanent feature of the universe as matter.

This conclusion has not been lost on Fred Hoyle, whose book *The Intelligent Universe* describes a state of affairs very much like the one I have just outlined.[4] Unless some law of nature limits the growth of intelligence and technology, or forbids intelligent life forms from spreading across the universe though permitting simple organisms to do so, it is hard to see how Hoyle's dramatic proposals can be avoided. Francis Crick and Leslie Orgel have arrived at a similar conclusion. Impressed by the substantial difficulties that scientists face in explaining biogenesis, they proposed the idea of "directed panspermia," according to which Earth was deliberately seeded with life by intelligent aliens.[5] By extension, life could be spread around the entire universe this way, without having originated anywhere in particular.

Many people find the idea of universal life very attractive. Scientifically, however, it seems a bit of a cheat. It tries to dodge the problem of life's origin by shifting the problem off into space and back in time until it disappears entirely from view. Although there is nothing logically wrong with the theory that life and the universe have always existed, it offers no explanation for either. You don't explain something simply by declaring that it has always been there. So, from now on, I shall assume that life did begin somewhere and somehow, perhaps independently in many places, and ask what this implies for the nature of the universe.

Are the laws of nature rigged in favor of life?

> *The universe was not pregnant with life, nor the biosphere with man.*
>
> JACQUES MONOD[6]

> *You are wrong. They were.*
>
> CHRISTIAN DE DUVE[7]

Jacques Monod pointed out that everything in nature is the product of two fundamental factors: chance and law—or necessity, as he chose to call it. Take, for example, the orbit of the Earth around the Sun. Its elliptical form follows from Newton's laws of motion and gravitation. We might say that the shape of the orbit is necessarily elliptical. On the other hand, the specific size of the orbit—i.e., how far away Earth is on average from the Sun—is the product of many complicated factors, including some historical accidents related to what hit what in the solar nebula. There is no necessity for the Earth to orbit 150 million kilometers from the Sun, as opposed to, say, 200 million kilometers. The actual orbit is therefore part necessity, part chance. If we find an Earth-like planet in another star system, it will not match our own orbit kilometer for kilometer, but the law of gravitation will require it to follow an elliptical path.

An extreme example of necessity is the structure of a crystal.

The geometrical arrangement of a crystal lattice is determined entirely by the interatomic forces operating. Two pure salt crystals will have identical crystalline structures, as will two diamonds. Chance doesn't enter into it: crystals necessarily have the form that they do. By contrast, an extreme example of chance is the pinball machine. Certainly the ball obeys Newton's laws of motion between collisions with the pins, but its final destination is purely accidental. We do not expect pinballs to end up always in the same hole.

When it comes to life, how much is due to chance and how much to necessity? Monod himself was in no doubt. It was overwhelmingly the product of chance, he maintained, a perspective extolled in his famous book *Chance and Necessity*. Moreover, the chanciness of life applied, Monod contended, not only to the random and directionless nature of evolution, but to the physical processes that produced life in the first place. For Monod, the genesis of life was just a quirk of fate, the result of a blind cosmic lottery. As I explained in chapter 3, the probability that life formed solely by random molecular shuffling alone is infinitesimal. If that is how it happened, it will have happened only once in the observable universe.

If life is discovered on Mars or elsewhere, and if we can be sure that a panspermia process is *not* involved, then Monod's doctrine of chance, and the bleak, heroic philosophy that goes with it, will be discredited. Those who believe that we are not alone in the universe already reject blind chance as an explanation for the origin of life. They suppose that an element of necessity, or lawfulness, is involved. In other words, they assume that the emergence of life from nonliving chemicals is the result of the normal outworking of universal laws, and that if those laws work themselves out by producing life here on Earth they will in all likelihood produce life on other planets too. It is a point of view clearly stated, for example, by the Space Science Board of the U.S. National Academy of Sciences in its assessment of the potential for life on Mars: "Given that life arose on Earth, it seems possible or even plausible that life could have arisen on Mars under similar conditions and at much the same time."[8]

The belief that, since life exists on Earth, it must be common throughout the universe, is sometimes called biological determinism or predestination.[9] It seems to be widespread among astronomers, chemists, and physicists, but much rarer among biologists. In weighing the relative importance of chance and necessity in the origin of life, most biologists come down on Monod's side in favor of chance as the dominant factor. But there are exceptions. Christian de Duve, like Monod a Nobel Prize winner, thinks that the formation of life is inevitable and swift under the right conditions. His recent book *Vital Dust* has the subtitle *Life as a Cosmic Imperative*. De Duve believes the universe to be a "hotbed" of life, which emerges as an automatic consequence of the laws of nature. "Life is the product of deterministic forces," he writes. "Life was bound to arise under the prevailing conditions, and it will arise similarly wherever and whenever the same conditions obtain. . . . Life and mind emerge not as the result of freakish accidents, but as natural manifestations of matter, written into the fabric of the universe."[10]

What, then, are these bio-friendly laws that apparently encourage disordered matter and energy to fast-track along the path to life? Is there some special biological principle at work, or will the ordinary laws of physics do the trick? Historically, both points of view have been held. Aristotle, for example, proposed that life is the manifestation of a universal organizing principle. Darwin too suggested, "The principle of life will hereafter be shown to be a part, or consequence, of some general law."[11] I think it is fair to say, however, that few biologists today believe there are laws of life in quite the same way as there are laws of physics. Many find the idea of special laws or principles to guide the development of matter towards life, over and above the basic laws of physics, altogether too mystical, too reminiscent of vitalism.

So perhaps the necessary powers to produce life are already implicit in the laws of physics themselves? Imagine life emerging from a soup in the same dependable way that a crystal emerges from a saturated solution, with its final form predetermined by the interatomic forces. Consider, for example, the way amino acids link together to form polypeptides, the stuff of proteins. To have biological function, the amino acids must be joined in a suitable sequence. If they connect up in any old permutation, the chances of getting a useful protein are negligible. But suppose the interatomic forces that operate to forge the peptide bonds can discriminate between different sequences? Maybe these forces will prefer to join the amino acids in combinations that happen to be biologically helpful.

Occasionally researchers claim precisely this. Gary Steinman and Marian Cole, working at Pennsylvania State University in the 1960s, tested reports that amino acids might form peptide chains in a manner that was "anything but random."[12] Their experiments seemed to confirm that molecules significant for life are made preferentially. "These results prompt the speculation that unique, biologically pertinent peptide sequences may have been produced prebiotically," they wrote. Steinman and Cole also noted that "preferential interaction has been observed at higher levels of organization as well," going so far as to allege that "a type of built-in 'predestination' can be identified at several levels of biological order."

Steinman and Cole imply that matter has an innate tendency to grope in the direction of life by virtue of the chemical affinities that act between atoms and molecules. They are not alone. Sidney Fox also concludes that "amino acids determine their own order in condensation,"[13] and that this nonrandom "self-instruction" infuses macromolecules with crucial biological information, paving the way for life. The late Cyril Ponnamperuma, who, like Sidney Fox, was one of the early pioneers in biogenesis research, believed that "there are inherent properties in the atoms and molecules which seem to direct the synthesis" towards life.[14] Ponnamperuma repeats the familiar line of reasoning that, because the building blocks of life are widespread in the universe, life should be too. "Radio astronomers have discovered a vast array of organic molecules in the interstellar medium. We are thus led to the inescapable conclusion that life must be commonplace in the cosmos."[15] (On page 90, I showed that this argument is totally bogus; to revisit my metaphor, bricks alone don't make a house.)

If we envisage a soup of chemicals, and the near-infinite range of possible reactions, there will be a vast decision tree of molecular arrangements open to it. Only a few tiny twiglets on the tree will lead to life. Fox and Ponnamperuma suggest that preferential chemical affinities serve to entice the participating molecules along the appropriate pathway through this tree until life is attained. If this were true, it would be astounding, not to say incredible. To claim that *atomic* processes include a built-in bias favoring *organisms* means that the laws of atomic physics effectively contain a blueprint for life. There would be a link between the basic forces that act on atoms, and the final complicated macroscopic product—a functioning organism. But what would be the nature of that link? How can the basic laws of physics "know" about complex, information-laden entities like living cells?

The heart of my objection is this: The laws of physics that operate between atoms and molecules are, almost by definition, simple and general. We would not expect them alone to lead inexorably to something both highly complex and highly specific. Let me try to spell out where the problem lies. In chapter 4, I pointed out that genomes are more or less random sequences of base pairs, and that this very randomness is essential if they are to play the role of evolvable, information-rich molecules. But this fact flatly contradicts the claim that genes can be generated by a simple, predictable, lawlike process. As I explained in that chapter, a law is a way to compress data algorithmically, to boil down apparent complexity to a simple formula or procedure. Conversely, no simple law can generate, alone, a random information-rich macromolecule to order. A law of nature of the sort that we know and love will not create biological information, or indeed any information at all. Ordinary laws just transform input data into output data. They can shuffle information about but they can't create it. The laws of physics, which determine what atoms react with what, and how, are algorithmically very simple; they themselves contain relatively little information. Consequently they cannot on their own be responsible for creating informational macromolecules. Contrary to the oft-repeated claim, then, life cannot be "written into" the laws of physics—at least, not into anything like the laws of physics that we know at present.

If we accept that the genome is random and information-rich, appealing to nonrandom chemistry to make life is a clear contradiction. Nonrandomness is the exact *opposite* of what is needed to produce a random macromolecule. The whole point of the genetic code, for example, is to *free* life from the shackles of nonrandom chemical bonding. A genome can choose whichever amino-acid sequence it wants, regardless of the chemical preferences of molecules. It achieves this by deploying special enzymes designed precisely to override the nonrandom tendencies of chemistry. That is why life goes to all the trouble of having coded information and software-mediated assembly, via the nucleic-acid/protein contract. Life works its magic not by bowing to the directionality of chemistry, but by *circumventing* what is chemically and thermodynamically "natural."

Of course, organisms must comply with the laws of physics and chemistry, but these laws are only incidental to biology. Their main role is to permit an appropriate logical and informational system to come into being. Where chemical reactions are easy and thermodynamically favored, life will cheerfully make use of them, but if life needs to perform "unnatural" chemistry, it finds a way. It fabricates the necessary catalysts to make weird reactions go, and manufactures appropriate energized molecules, sometimes in complicated combination, to drive against thermodynamic gradients. The key step that was taken on the road to biogenesis was the transition from a state in which molecules slavishly follow mundane chemical pathways, to one in which they organize themselves to forge their own pathways. The chalk-and-cheese mixing ability of software control, as exemplified in the use of a genetic code, is the clearest manifestation of this transcendence. Life opts out of the strictures of chemistry by employing an information-control channel, freeing it to soar above the clodlike blunderings of atomic interactions and create a new, emergent world of autonomous agency.

Once this essential point is grasped, the real problem of biogenesis is clear. Since the heady successes of molecular biology, most investigators have sought the secret of life in the physics and chemistry of molecules. But they will look in vain for conventional physics and chemistry to explain life, for that is a classic case of confusing the medium with the message. The secret of life lies, not in its chemical basis, but in the logical and informational rules it exploits. Life succeeds precisely because it *evades* chemical imperatives.

There is, I should mention, a curious loophole in my argument. Recall the discussion in chapter 4 about algorithmic complexity and binary sequences. If you find a compact formula that generates a given sequence, you have obviously proved that the sequence isn't random. However, if you try to find a formula and fail, you haven't proved the opposite—that the sequence definitely *is* random. It may be that you have overlooked a very obscure formula that will generate a given random-looking sequence. In fact, it can be shown that it generally isn't possible to prove randomness.[16] Translated into the problem of biogenesis, this means we can never rule out the possibility that a genome has been generated in a simple lawlike manner—for example, by cleverly rigged laws of physics. But there is a price to pay. If the above were so, it would mean that life only *seems* complicated but is actually very simple.

There are many instances in nature of deceptively complicated systems. Patterns that form spontaneously may look complex to the casual eye but turn out to have a hidden, underlying simplicity. Examples of this include shapes that display enormous wiggliness or intricacy, such as coastlines, the surfaces of slippery sand piles, and the rings of Saturn. Many natural features of this sort can be accurately modeled by a type of geometrical object known as a fractal. Fractals look infinitely irregular and complex, but in fact they possess a simplifying mathematical property called self-similarity. Roughly speaking, a self-similar pattern is one in which the degree of irregularity is the same on all length scales. As a result, the description or generation of fractals does not, in fact, require a great deal of information.[17] One of the most famous fractals, the Mandelbrot set, which is often displayed in color as an art form, can be created on a computer by an extremely simple algorithm.[18] Thus many non-biological systems that look like examples of random complexity are in fact highly nonrandom after all.

Could life be like this: apparently complex but actually very simple, like a fractal, and therefore the product of a simple lawlike process? It isn't necessary to suppose that all life is simple: only the first living thing. Once life was goaded into existence by a law, Darwinian evolution could then add irreducible complexity. Personally I do not believe it, not least because it demands a view of nature that is incredibly contrived. To claim that there really is "a code within the code," generating living creatures on demand from simple formulae, is just too far-fetched.

Is it Darwinism all the way down?

In the previous section I argued that, barring a cunning setup in which life is actually simplicity masquerading as complexity, normal physical laws alone can't crank out life to order. Yet this doesn't mean all forms of biological determinism are doomed from the outset. It could still be that life is inevitable, or at least strongly favored, given appropriate conditions. Some scientists suggest a weaker, and more credible, form of biological determinism. Christian de Duve, for example, sees chance playing a role, but chance tempered by various physical constraints that impose an overall directionality—with life as the predictable destination. These constraints, though stringent, are not so specific as to dictate the precise details of the chemical synthesis. Rather, de Duve likens the situation to water flowing obligatorily from a crater into a gorge, its general direction predetermined by the conformation of the landscape. Thus he feels able to write: "The emergence of life was the outcome of highly deterministic processes, virtually bound to occur under the physical–chemical conditions that prevailed at the time."[19]

There are also the ideas of Stuart Kauffman, which I discussed in chapter 5. Kauffman doesn't claim that there is a pre-existing blueprint for life, only a propensity for organized complexity to emerge under suitable conditions. So life may not be such a surprise after all: "An expected collective property of complex systems" is the way he expresses it.[20] He thinks, "The routes to life are many and its origin is profound yet simple." According to Kauffman's theory, there is no *specific* end goal encoded in the principles of self-organization, no earmarked microbe, only a general trend towards the sort of complex states that are likely to lead to life.

Attractive though these arguments may be, we are still left with the mystery of where biological information comes from. The objections I gave in the previous section remain valid. If the normal laws of physics can't inject information, and if we are ruling out miracles, then how can life be predetermined and inevitable rather than a freak accident? How is it possible to generate random complexity and specificity *together* in a lawlike manner? We always come back to that basic paradox.

There *is* a solution to this problem, I think, but it is a radical one that many scientists are extremely reluctant to contemplate. Yet, the more I puzzle over the problem of biogenesis, the more I feel that we cannot escape embracing something like it. Let me give a sketch of what I have in mind. In chapter 2, I mentioned that Schrödinger was sufficiently puzzled by life to suggest "a new type of physical law." I think he was on the right track. However, we don't need another law of physics. We must look elsewhere. But where?

Two fields of inquiry offer tantalizing clues. The first is complexity theory. I have already mentioned Kauffman's related ideas on chemical networks and autocatalytic cycles. In recent years a lot of work has been done on the study of complex systems in general. Many investigators have come to the conclusion that there are universal mathematical principles governing the way that such systems behave. These "laws" cannot be derived from the underlying laws of physics, because they are not physical laws in the usual sense. Instead, they arise from the logical structure of the system, and depend only indirectly on the physical forces involved. For this reason such systems can readily be modeled as "games" on computers. Many such computer models display strikingly lifelike qualities; one is even called *The Game of Life*.[21] There is now an expanding field of research known as "artificial life" based on such computer models.[22] The hope of many complexity theorists is that some sort of self-organizing physical processes could raise a physical system above a certain threshold of complexity at which point these new-style "complexity laws" would start to manifest themselves, bestowing upon the system an unexpected effectiveness to self-organize and self-complexify. The result would be a series of transitions that ratchet the system abruptly up the complexity ladder. Under the bidding of such laws, the system might be rapidly directed towards life. If that is correct, it would mean that life is not so much written into the laws of physics as built into the logic of the universe.

My own opinion is that emergent laws of complexity offer reasonable hope for a better understanding not only of biogenesis, but of biological evolution too. Such laws might differ from the familiar laws of physics in a fundamental and important respect. Whereas the laws of physics merely shuffle information around, a complexity law might actually *create* information, or at least wrest it from the environment and etch it onto a material structure.[23] This would represent a major departure from the traditional reductionist picture of the world, in which forces act between inert particles of matter, and information is treated as a secondary, derivative concept. My proposal means accepting that information is a genuine physical quantity that can be traded by "informational forces" in the same way that matter can be moved around by physical forces. It also means accepting complexity as a physical variable, with real causal efficacy, rather than a merely qualitative description of how complicated a system is. I believe it is only under the action of an informational law that the information channel, or software control, associated with the genetic code could have come into existence (see page 115).

I may have made my proposal seem more radical than it is. The idea of informational, or software, laws isn't all that new. Many other investigators have suggested something similar. For example, Manfred Eigen has written, "Our task is to find an algorithm, a natural law that leads to the origin of information."[24] Though acknowledging the crucial role played by molecular Darwinism, Eigen and his colleagues nevertheless see the need for it to be augmented by other physical processes which can be an additional source of biological information.[25]

I first mooted the idea of "software laws" some years ago in my book *The Cosmic Blueprint*. There, I envisaged the new laws as consistent with, but not reducible to, the underlying laws of physics. When I set out to write the present book, I did not think that such laws were necessary to explain biogenesis. Instead, I assumed it was a case of "Darwinism all the way down." Impressed by the laboratory work on the fabrication of replicator molecules, and the apparent ease with which simple organic building blocks can form, I found it plausible that chance alone could produce a small replicator molecule rather quickly. After that, molecular evolution would take over, driving the system steadily towards cellular life. Having studied the many variants of that theory on offer, I am now much more skeptical. It seems to me very unlikely that all that is necessary is for the right chemical reaction or the right molecule to turn up. Real progress with the mystery of biogenesis will be made, I believe, not through exotic chemistry, but from something conceptually new.

A blend of molecular Darwinism and laws of organizational complexity could offer a way forward. In such a scenario, relatively small replicator molecules form by chance and start to evolve by Darwinian means, but the process is sometimes aided, and even overridden, by organizational principles that confer specificity and information.[26] These organizational principles serve to amplify greatly the selectivity of the evolutionary process, and lead to sudden jumps in complexity rather than the incremental advance expected from Darwinian evolution acting alone.

The second line of inquiry that may or may not have a bearing on biogenesis is rather more speculative. It involves quantum mechanics, the theory that describes the weird behavior of matter at the atomic level. Mostly, biochemists and molecular biologists ignore quantum mechanics. Atoms and molecules are treated like little building blocks that stick together in various shapes, but the reality of the microworld is far more subtle than that. For a start, there is the famous wave-particle duality: an atom has both wavelike and particlelike aspects. Significantly, the wave can be identified with information or software, because it describes what is known about the system. On the other hand, the atom treated as a particle corresponds to hardware. When a quantum measurement is made, the wave "collapses"—changes suddenly—because the knowledge of the system changes. But this in turn affects the subsequent behavior of the particle.[27] There is thus a sort of hardware-software entanglement in quantum mechanics. Information (or knowledge) has downward causative power. So here is a mainstream physical theory that has information at its heart, which it tangles with matter in an intimate way. Furthermore, the interatomic forces that form biological molecules like proteins and nucleic acids are indeed quantum-mechanical in nature. Could some sort of quantum-organizing process be just what is needed to explain the origin of informational macromolecules?

Supporting evidence for this conjecture comes from an unusual direction. In his famous book, Erwin Schrödinger proposed that the unit of heredity is "an aperiodic crystal." By this he meant a molecular structure stable enough to retain its form, but complex enough to store a lot of information. A normal periodic crystal has stability but low algorithmic-information content (see page 116). Schrödinger's idea proved prophetic. A DNA molecule has structural stability (though it's not perfect—the preservation of information requires the use of proofreading and editing processes). The aperiodicity arises because the sequence of bases is mostly random, and hence information-rich—a point I have belabored.

A few years ago, chemists were startled by the discovery of a rather different sort of aperiodic crystal, termed a quasi-crystal. Quasi-crystals possess a curious fivefold symmetry; that is, they look the same when rotated through seventy-two degrees. However, unlike normal crystals, they are not periodic. Indeed, it can be proved that the pattern of atoms never repeats itself.

The reason that quasi-crystals came as a surprise goes back to simple geometry. It is well known that you can tile a wall with triangles, squares, and hexagons, but not with pentagons. Pentagons do not tessellate—they leave gaps. So fivefold symmetry will not permit a simple repetitive pattern. However, in a famous theorem, Roger Penrose proved that an infinite wall can be tessellated with fivefold symmetry using two differently shaped tiles—a fat and a thin rhombus.[28] Quasi-crystals are a naturally occurring three-dimensional analogue of a Penrose tiling pattern. Penrose himself has suggested that the very existence of quasi-crystals presents a puzzle, in view of its aperiodic nature. A normal periodic crystal can grow atom by atom, because it forms a regular repeating structure, but a quasi-crystal requires some sort of long-range organization to make sure that the right bits fit in the right places. Penrose thinks that subtle aspects of quantum mechanics, and even quantum gravity, may play a role in this geometric organization.

Because of its fivefold symmetry, a quasi-crystal has very little information stored in its orientation, but an unlimited amount in its linear aperiodic sequence. It thus combines something of Cairns-Smith's idea of impure crystals, and something of Schrödinger's idea of an aperiodic chain molecule. Like DNA, quasi-crystals seem at first sight to be "impossible objects," with enormous algorithmic complexity. Yet somehow quantum mechanics permits them to come into existence. I am not suggesting that quasi-crystals are possible genomes (though who knows?), only that their study may elucidate how quantum mechanics can organize the formation of complex physical structures with high information-storage capacity.[29]

A further hint that quantum magic might be afoot in the husbanding of biological information comes from the fashionable study of quantum computation.[30] It has been shown that a quantum computer can render some computationally intractable problems tractable (e.g., the traveling-salesman problem that I mentioned on page 121), again suggesting that a computationally "impossible object," such as an algorithmically random genome, might be produced rather readily by quantum processes, even though it would require a long and tortuous evolution by classical means.

I concede that the ideas I have skimmed over in this section are highly conjectural, but the very fact that the problem of biogenesis prompts such speculation underscores just how stubborn a mystery it is. Nevertheless, the assumption that life is a fundamental cosmic phenomenon, predestined to develop whenever conditions permit, remains widespread. Few proponents of the "life-will-out" thesis fully appreciate the sweeping implications of what they are proposing. Deterministic thinking, even in the weaker forms of de Duve and Kauffman, represents a fundamental challenge to the existing scientific paradigm. It is, in fact, enough to make most biologists shudder. Although biological determinists strongly deny that there is any actual design, or preordained goal, involved in their proposals, the idea that the laws of nature may be slanted towards life, even if not contradicting the letter of Darwinism, certainly offends its spirit. It slips an element of teleology back into nature, a century and a half after Darwin banished it. For many scientists, biological determinism is tantamount to a miracle in nature's clothing. That, of course, doesn't make it wrong. It might still be true! Life might indeed be bound to occur whenever conditions are suitable. But if that is so, the consequences will be profound indeed.

For three hundred years, science has based itself on reductionism and materialism, leading inevitably to atheism and a belief in the meaninglessness of physical existence. A bio-friendly universe would mark a decisive shift. The momentous significance has been eloquently expressed by de Duve: "From the perspective of determinism . . . I view this universe not as a 'cosmic joke,' but as a meaningful entity—made in such a way as to generate life and mind, bound to give birth to thinking beings able to discern truth, apprehend beauty, feel love, yearn after goodness, define evil, experience mystery."[31]

A ladder of progress?

In the history of science, no idea struck more deeply at mankind's self-esteem than Darwin's theory of evolution. The very public clash between Darwin and the Christian Church provides a classic example of how painful it can be when scientific developments fundamentally change the conceptual basis on which we build our theories of nature. Today evolution is almost universally accepted; even the Pope has given it his blessing. Yet in the quiet halls of academe a shadow of the old battle is still being fought. It hasn't attracted so much attention, and few theologians have joined in, but in terms of its philosophical significance this latter-day skirmish is as important as the nineteenth-century struggle between Darwin and Wilberforce.

The issue at stake today is not whether life has gradually evolved over billions of years—the evidence for that is overwhelming—but whether there is something slanted about the manner of that evolution. In the nineteenth century, it was fashionable to regard life as developing along an upward path. Primitive life, it was said, slowly "improved" and changed into ever more elaborate and sophisticated forms, culminating in *Homo sapiens*, with our much-vaunted intelligence and powers of reasoning. Viewed this way, evolution was not so much a meandering path as a ladder of progress, leading steadily upwards from microbes to man. To be sure, the climb up that ladder has been brutal and wasteful, as natural selection took its toll, but in that progressive trend there was an austere glory and a special status for mankind.

The image of an evolutionary ladder of progress remains a potent symbol, and is still carried subconsciously by many scientists and laymen who do not appreciate the profound metaphysical assumptions that go with it. If evolution really is progressive, the laws of nature might not only be rigged in favor of creating life, but rigged in favor of advancing it too.

Opponents of "progressive" biology slam the idea on several

grounds. First, they point out, it implies a value judgment, that humans are somehow "better" than apes or frogs. Terms like "higher" mammals or "lower" vertebrates, which reflect traditional ladder-of-progress thought, betray this bias and are regarded as politically incorrect. Just what is it about humans, ask the critics, that make them an improvement on other organisms? In terms of sheer numbers, microbes win hands down. If adaptational success is the key criterion, superbugs are pretty adept at coping with environmental stress. Humans, of course, have high intelligence. That makes us successful when it comes to IQ, but we are hopeless swimmers and we can't fly. If we decide that intelligence is what matters, we can undeniably place ourselves at the top of the ladder. But is this not simply a case of chauvinism? We ourselves have selected the criterion that makes us top. We have decided our favored place and erected a ladder beneath us. It is no surprise, if we look back down, that the lower rungs are filled with less intelligent precursors. But so what? Is intelligence better in any absolute sense than, say, eyesight or hearing, both of which are only moderately well developed in humans?

These difficulties have made "progress" an unacceptable word for biologists. Nevertheless, there may still be some property of organisms—a quality of a more culturally neutral character—that might display a general "upward" trend over time. It is often suggested that complexity is such a property. The biosphere as a whole is undeniably far more complex today than it was three billion years ago. Also, the most complex organisms today clearly have a much greater complexity than the most complex organisms in the remote past. True, it has not been an entirely unbroken upward march. From time to time, catastrophic annihilation has occurred, perhaps due to asteroid impacts, resulting in the sudden elimination of the majority of species across the planet. These episodes certainly reduced biological complexity dramatically. But (so far) it has always bounced back with renewed vigor. The impression we gain is that life, *when left to flourish*, rides an escalator of growth, filling out every available niche, exploring new and better possibilities, developing ever more elaborate forms.

This systematic advance in organized complexity is so striking it has the appearance of a law of nature. It rests comfortably with recent cosmological thinking, which sees the universe as a whole increasing in complexity since the big bang. A more careful assessment, however, uncovers serious problems with this simple picture.

First, the principles of Darwinism rule out the teleological notion that life strives for betterment. Darwinian evolution works by applying the filter of natural selection to blind variation on a moment-by-moment basis, locking in the good changes and rejecting the bad. There is no mechanism within this paradigm for foresight, no way that a systematic march towards a predetermined goal could be set in train. If greater complexity makes good survival sense at the time, and only at the time, it gets selected. If not, it gets rejected.

Second, many organisms have grown less complex with time, such as fish that dwell in dark caves and have lost the use of their eyes. This is no surprise. There can be circumstances in which too much complexity is a positive nuisance. Redundant organs may hinder survival under ascetic conditions, or prove to be excess baggage when the going is good. A classic example of biological regress is Spiegelman's monster that I discussed in chapter 5. There the spoon-fed RNA slimmed itself down to a fraction of its original viral size in order to replicate faster.

When the fossil record is examined, the data generally support the contention that overall biological complexity increases with time. Some species grow simpler; others become more complex. But, barring global catastrophes, the average goes up. However, we must be careful about the notion of average. Life started out with simple microbes. If it was to go anywhere, that would inevitably be in the direction of greater complexity. According to Darwinism, evolution has the character of a random walk through the realm of biological possibilities, a blind, undirected groping. Obviously, if you start from a specially simple initial state, even a random excursion will likely take you in the direction of greater complexity, at least at first.

Stephen Jay Gould has explained this point well by using the analogy of a drunk leaning against a brick wall who then begins stumbling about blindly and eventually ends up falling in the gutter.[32] The drunk reaches the gutter not because he is seeking it out and methodically moving towards it. He is, in fact, staggering randomly: at any given time, the drunk is as likely to be moving towards the wall as away from it. The point is that, because the wall bounds his motion in one direction, he is obviously on average likely to be found somewhat away from the wall, and in due course he is going to encounter the gutter simply by chance. Gould points out that there is a limit to the degree of simplicity an organism may have and still be termed alive; this corresponds to the wall. If life on Earth began "at the wall"—i.e., with the simplest cells—and then evolved at random, the average complexity would inevitably increase as the distribution spread out in a lopsided way; see figure 10.1(a). But Gould cautions us against interpreting this simple diffusion as a systematic trend. He asserts it is nothing more than a random exploration of available possibilities.

I think that Gould is completely correct. If the increase in complexity with time is merely the result of a random walk away from simplicity, it cannot be considered as a lawlike directionality. To qualify as a genuine trend, the data would have to resemble figure 10.1(b). Whether there *is* a real trend in evolution in addition to a drunken walk is a matter for scientific inquiry to decide. So what are the facts? Is it (a) or (b)?

Unfortunately, the situation is not easy to investigate. Larger, more complex organisms tend to be noticeable, so we accord them a status lacked by the microbes. But as Gould has emphasized, most life on Earth is microbial. So-called advanced life is really a tail in the distribution, and we must be careful not to let the tail wag the dog. On the other hand, microbiologists believe that even microbes are reasonably highly evolved. Undoubtedly the most "primitive" microbe today is still far more complex than the first living cell. So, although most life on Earth is "stuck" at the level of microbes, even within this class there would seem to have been a general trend towards complexity. When it comes to multicellular life, the most di-

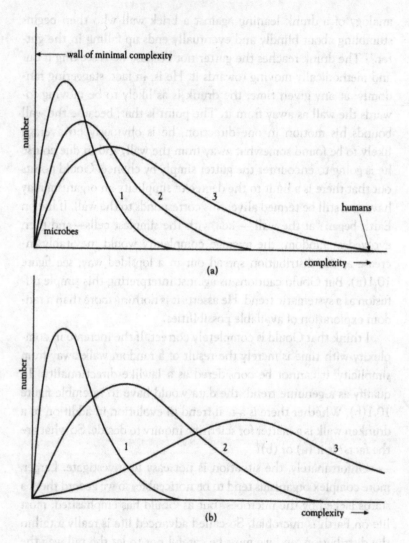

Figure 10.1. Ladder of progress? Biological complexity increases over time, but is there a systematic trend, or just a random diffusion away from a "wall of simplicity"? The diffusion model, supported by Gould, is shown in (a). Curves 1, 2, and 3 represent successive epochs. Life remains dominated by simple microbes, but the tail of the distribution edges to the right. If there were a definite drive towards complexity, the curves would look more like those in (b).

rect way to check—an examination of the fossil record—is unfortunately rather ambiguous. The record itself is fragmentary. There is, for example, a trend towards greater complexity in the acceleration of primate brain size with time. Set against this, Gould cites the work of Dan McShea of the Santa Fe Institute for the Study of Complexity, which fails to find any general trend towards complexity in the vertebral column.[33] On the whole, the evidence for a systematic advance of complexity is at best scrappy. The jury is still out.

Of course, the drunken-walk analogy is relevant only to the extent that evolution is expected to be random. Richard Dawkins has stressed that, although individual mutations are generally random, natural selection is anything but.[34] Selection filters out those organisms less favorably adapted to their circumstances and rewards those that are better adapted, which inevitably drives a trend in the direction of better *adaptation*. But better adaptation may or may not involve increasing complexity. The definition of "best-adapted organism" will vary anyway, depending on changing environmental circumstances. There is no preordained "best fit," no optimal adaptation, and no fixed "goal" towards which natural selection steers evolution. Any directionality in adaptation is likely to involve a process of temporary adjustment, and not form part of an overall trend.

Most biologists are of the opinion that any growth of complexity can be satisfactorily accounted for by the drunken-walk effect. There are suspicions, however, of a hidden ideological agenda. Gould makes no secret that he thinks complexity is being used as a surrogate for progress, which he regards as a "noxious" concept on ideological grounds. Thus he writes: "I believe that the most knowledgeable students of life's history have always sensed the failure of the fossil record to supply the most desired ingredient of Western comfort: a clear signal of progress measured as some form of steadily increasing complexity for life as a whole through time."[35] But Gould has no truck with this cozy view, and sees ironic splendor in life's very pointlessness: "We are glorious accidents of an unpredictable process with no drive to complexity, not the expected results of evolutionary principles that yearn to produce a creature capable of understanding the mode of its own necessary construction."[36] Belief in advancing complexity is, according to Gould, a nostalgic relic of pre-Darwinian sentimentality, with its muddled ideas of supernatural design. Having banished the guiding hand of God from the biosphere a century and a half ago, biologists are understandably reluctant to let it back in in the guise of a law of nature.

Again, I agree with Gould. A trend of increasing complexity *would* provide evidence of purpose in the universe. Of course, if there were such a trend, it would not preclude a major role for chance too. The question would then arise precisely what biological features are the result of chance and what might be expected as part of a trend. It is hard to imagine that minor details, such as the number of digits or the existence of eyebrows, could be the direct manifestation of a fundamental law. On the other hand, the essential architecture of multicellular organisms might well be the product of certain mathematical principles of organization. In my opinion that will turn out to be the case. But I shall let de Duve have the last word on this. He suggests that, once the tree of life is shorn of its elaborate canopy of foliage, "the structure of the trunk, with its progressive rise toward greater complexity, is clearly evident."[37]

Is mind predestined?

The universe has invented a way to know itself.
ALAN DRESSLER[38]

Of all the complex structures produced by terrestrial biology, none is more significant than the brain, the most complex organ of all. Are brains just random accidents of evolution, or are they the inevitable by-products of a lawlike complexifying process? There is a commonly held assumption that if life arises on other planets it will parallel life on Earth in its evolutionary development. Supporters of

SETI, the search for extraterrestrial intelligence, argue that over billions of years extraterrestrial life will complexify to form plants and animals, and eventually discover cognition and intelligence, exactly as happened here. On at least a fraction of inhabited planets, they maintain, intelligent life will advance to the point of technology, and some of those technological communities may even now be trying to contact us with radio signals. Thus SETI researchers generally subscribe to the ladder-of-progress concept, accepting that not only life but also mind is in some sense predestined to arise in the universe.

This viewpoint, though prevalent, again conceals a huge assumption about the nature of the universe. It means accepting, in effect, that the laws of nature are rigged not only in favor of complexity, or just in favor of life, but also in favor of mind. To put it dramatically, it implies that mind is written into the laws of nature in a fundamental way. It is therefore highly significant, surely, that the products of nature's complexifying trend—intelligent beings like *Homo sapiens*—are able to understand the very laws that have given rise to "understanding" in the first place.

It is an inspiring vision. But is it credible? Can we believe that the universe is not only bio-friendly but mind-friendly too? In 1964, the biologist George Simpson wrote a skeptical paper entitled "On the Nonprevalence of Humanoids,"[39] in which he emphasized the futility of the search for advanced extraterrestrial life. He termed it "a gamble of the most adverse odds in history." Pointing out that humans are the product of countless special historical accidents, he concluded: "The assumption, so freely made by astronomers, physicists, and some biochemists, that once life gets started anywhere, humanoids will eventually and inevitably appear is plainly false." In a recent debate with SETI supporter Carl Sagan, the biologist Ernst Mayr echoed Simpson's skepticism: "On Earth, among millions of lineages or organisms and perhaps 50 billion speciation events, only one led to high intelligence; this makes me believe its utter improbability."[40]

Stephen Jay Gould similarly denounces the notion that life is destined to produce mind. Imagine, he says, some catastrophe that wiped out all advanced life on Earth, leaving only microbes. If the evolutionary drama were rerun, what would happen? Would we expect a broadly similar pattern of development, with fish, vertebrates, reptiles, mammals, and intelligent bipeds re-emerging? Not a bit of it, he concludes. The history of life on Earth is a gigantic lottery, with far more losers than winners. It contains so many accidents of fate, so many arbitrary quirks, that the pattern of change is essentially random. The millions of fortuitous steps that make up our own evolutionary history would surely never happen the second time around, even in broad outline. History would "cascade down another pathway," so that "the vast majority of replays would never produce . . . a creature with self-consciousness. . . . The chance that this alternative [i.e., rerun] set will contain anything remotely like a human being must be effectively nil."[41]

It is hard to fault the logic of Simpson and Gould's argument. If evolution is nothing but a lottery, a drunken walk, there is little reason why life should go beyond the level of microbes, and no expectation whatever that it will advance obligingly towards intelligence and consciousness, still less develop humanoid characteristics. We should then be forced to agree with Monod's melancholy conclusion: "Man at last knows he is alone in the unfeeling immensity of the universe, out of which he has emerged only by chance."[42] Only if there is more to it than chance, if nature has an ingeniously built-in bias towards life and mind, would we expect to see anything like the developmental thrust that has occurred on Earth repeated on other planets.

The search for life elsewhere in the universe is therefore the testing ground for two diametrically opposed world-views. On one side is orthodox science, with its nihilistic philosophy of the pointless universe, of impersonal laws oblivious of ends, a cosmos in which life and mind, science and art, hope and fear are but fluky incidental embellishments on a tapestry of irreversible cosmic corruption. On the other, there is an alternative view, undeniably romantic but perhaps true nevertheless, the vision of a self-organizing and self-complexifying universe, governed by ingenious laws that encourage matter to evolve towards life and consciousness. A universe in which the emergence of thinking beings is a fundamental and integral part of the overall scheme of things. A universe in which we are not alone.

NOTES

PREFACE

1. Fred Hoyle, *The Black Cloud* (Harmondsworth: Penguin, 1960).
2. Eugene Wigner, "The Probability of the Existence of a Self-Reproducing Unit," in *The Logic of Personal Knowledge*, anonymous ed. (London: Routledge & Kegan Paul, 1961), p. 231.
3. Erwin Schrödinger, *What Is Life?* (Cambridge: Cambridge University Press, 1944).
4. Nevertheless, Carter's work has been widely discussed in the literature; see, for example, John Barrow and Frank Tipler, *The Anthropic Cosmological Principle* (Oxford: Clarendon Press, 1986).
5. B. Carr and M. J. Rees, "The Anthropic Principle and the Structure of the Physical World," *Nature* 278(1979):605.
6. See, for example, Manfred Eigen, *Steps Towards Life*, trans. P. Woolley (Oxford: Oxford University Press, 1992); Christian de Duve, *Vital Dust* (New York: Basic Books, 1995). The typical upbeat tone of much biogenesis comment can be discerned from the following quotation: "There is no reason to doubt that we shall discover, one by one, all the steps in physical and chemical evolution. We may even reproduce the intermediate steps in the laboratory. . . . We have the right to be quite optimistic." (Cyril Ponnamperuma, "The Origin, Evolution, and Distribution of Life in the Universe," in Clifford Matthews and Roy Abraham Varghese, eds., *Cosmic Beginnings and Human Ends* (Chicago: Open Court, 1993).

7. The term "god-of-the-gaps" is used by theologians to refer to attempts to explain gaps in the scientific understanding of nature by invoking selective divine action.

8. Quoted in Andrew Scott, *The Creation of Life* (Oxford: Blackwell, 1986), p. 49.

9. Paul Davies, *The Mind of God* (London: Viking; New York: Simon & Schuster, 1992).

10. Paul Davies, *Are We Alone?* (London: Penguin; New York: Basic Books, 1995).

CHAPTER 1: THE MEANING OF LIFE

1. Jacques Monod, *Chance and Necessity*, trans. A. Wainhouse (London: Collins, 1972), p. 167.

2. Francis Crick, *Life Itself: Its Nature and Origin* (New York: Simon & Schuster, 1981), p. 88.

3. David Mowaljarlai and Jutta Malnic, *Yorro Yorro* (Broome, Western Australia: Magabala Books, 1993), chap. 23.

4. The theory of common descent was actually first proposed by Charles Darwin's grandfather Erasmus Darwin, in *Zoonomia, or the Laws of Organic Life* (London, 1794).

5. *New Scientist*, February 10, 1996, p. 26.

6. Stephen Jay Gould, *Life's Grandeur* (New York: Random House, 1996), chap. 14. I shall return to this topic in chap. 10.

CHAPTER 2: AGAINST THE TIDE

1. H. W. Longfellow, "A Psalm of Life."

2. A. S. Eddington, *The Nature of the Physical World* (Cambridge: Cambridge University Press, 1928), p. 74.

3. A. I. Zotin, "The Second Law, Negentropy, Thermodynamics of Linear Processes," in I. Lamprecht and A. I. Zotin, eds., *Thermodynamics of Biological Processes* (New York: de Gruyter, 1978), p. 19.

4. A. S. Eddington, "The End of the World: From the Standpoint of Mathematical Physics," *Nature* 127(1931):447.

5. Erwin Schrödinger, *What Is Life?* (Cambridge: Cambridge University Press, 1944), p. 81.

6. Actually, the notion of comparing the entropy of two organisms is

rather vague. A more precise evaluation can be given in terms of the relative complexities of their genomes, expressed using a quantity called algorithmic entropy (see chapter 4). It is then the case that "higher" organisms have higher (not lower) algorithmic entropy, and so do not conflict with the second law anyway in this respect.

7. Original German text published in *Populare Schriften* (Leipzig) 26 (1905); English translation quoted in Klaus Mainzer, *Thinking in Complexity* (Berlin: Springer-Verlag, 1994), p. 82.

8. C. E. Shannon and W. Weaver, *The Mathematical Theory of Communication* (Urbana: University of Illinois Press, 1949).

9. The reader may be unsure about why making a selection is the same as introducing information, but in its broadest sense information is just the exclusion of possibilities. If a physical system has only one possible state, we learn nothing new by inspecting it. The more possibilities there are, the more we learn by discovering the actual state. Natural selection eliminates unfit organisms, and thus selects only certain genomes from among a much larger possible set. All the other possibilities are excluded. This amounts to adding information to the genomes.

10. D. S. Bendall, ed., *Evolution from Molecules to Men* (Cambridge: Cambridge University Press, 1983), p. 127.

11. The "wrong-way" property of gravitation is closely connected to the fact that gravitational energy is negative.

12. Roger Penrose, *The Emperor's New Mind* (Oxford: Oxford University Press, 1989), and *Shadows of the Mind* (Oxford: Oxford University Press, 1994).

13. Lee Smolin, *The Life of the Cosmos* (Oxford: Oxford University Press, 1997), p. 159.

14. For more about quantum nonlocality and its weird properties, see, for example, P. C. W. Davies and J. R. Brown, eds., *The Ghost in the Atom* (Cambridge: Cambridge University Press, 1986).

15. For details on how to set up such paradoxical situations, see Paul Davies, *About Time* (London: Viking; New York: Simon & Schuster, 1995), chap. 11.

CHAPTER 3: OUT OF THE SLIME

1. D. S. Bendall, ed., *Evolution from Molecules to Men* (Cambridge: Cambridge University Press, 1983), p. 128.

2. Adrian Desmond and John Moore, *Darwin* (London: Michael Joseph, 1991), p. 230. Note that the tree of life spreads out in the forward direction of time, in contrast to family trees, which (to start with, at least) spread out backwards in time.

3. See, for example, Gregory R. Bock and Jamie A. Goode, eds., *Evolution of Hydrothermal Ecosystems on Earth (and Mars?)* (New York: Wiley, 1996), chaps. 1, 2.

4. Modern stromatolites involve the activities of other micro-organisms too, as well as algae. It is hard to tell exactly what made the fossils.

5. For a recent review, see J. William Schopf, "The Oldest Known Records of Life: Early Archaean Stromatolites, Microfossils, and Organic Matter," in S. Bengtson, ed., *Early Life on Earth* (New York: Columbia University Press, 1994), p. 193.

6. S. J. Mojzsls et al., "Evidence for Life on Earth Before 3,800 Million Years Ago," *Nature* 384(1996):55.

7. Gerald Feinberg and Robert Shapiro, *Life Beyond Earth* (New York: William Morrow, 1980), p. 113.

8. Charles Thaxton, Walter Bradley, and Roger Olsen, *The Mystery of Life's Origin* (New York: Philosophical Library of New York, 1984), p. 12.

9. Bendall, ed., *Evolution from Molecules to Men*, p. 128.

10. Quoted in Andrew Scott, *The Creation of Life* (Oxford: Blackwell, 1986), p. 49.

11. Quoted in Svante Arrhenius, *Worlds in the Making* (London: Harper, 1908), p. 216.

12. If reactions take place on a surface, such as clay or rock, rather than in the body of a watery soup, the thermodynamic considerations are altered somewhat in favor of synthesis, since the molecules are confined to a two-dimensional domain.

13. This is much more than the number of atoms in the observable universe.

14. Fox himself claims that the right one is not achieved by accident, but that the chemistry itself favors the infinitesimal fraction of peptide chains that are biologically relevant. See, for example, S. Fox and K. Dose, eds., *Molecular Evolution and the Origin of Life* (New York: Marcel Dekker, 1977). The claim that chemistry somehow "knows about" biology is a sweeping and provocative one that I shall return to in chapter 10.

15. Fred Hoyle, *The Intelligent Universe* (London: Michael Joseph, 1983), p. 19.

16. *Omnia* (British Airways in-flight magazine), September–October 1997, p. 26.
17. An explanation that relies on freaky circumstances, although not impossible, is inherently implausible. We may take the odds against those circumstances as a quantitative measure of our disbelief, or lack of confidence, in the fluke theory.

CHAPTER 4: THE MESSAGE IN THE MACHINE

1. It is known that there are basic limits to what can be discovered about physical systems. For example, Heisenberg's uncertainty principle of quantum mechanics forbids knowledge of both the position and motion of an atom at the same time. There is thus a fundamental impenetrability to nature at the atomic level. Might the mystery of life also be impenetrable? The physicist Niels Bohr, one of the founders of quantum mechanics, once believed so. He concluded that life hides its secrets from us in the same way as an atom does. "On this view the existence of life must be considered as an elementary fact that cannot be explained," he wrote. See Niels Bohr, "Light and Life," *Nature* 131(1933):421, 457.
2. Occasionally a twenty-first amino acid is used.
3. John Maynard Smith and Eörs Szathmáry, *The Major Transitions in Evolution* (Oxford and New York: Freeman, 1995), p. 81.
4. Carl Woese, "Evolution of the Genetic Code," *Naturwissenschaften* 60(1973):447.
5. J. D. Bashford, I. Tsohantjis, and P. D. Jarvis, "A Supersymmetric Model for the Evolution of the Genetic Code," *Proceedings of the National Academy of Sciences (NY)* 95(1998):987.
6. Interestingly, there *are* apparently random sequences in DNA, known for obvious reasons as junk DNA, which seem to serve no crucial purpose.
7. See for example Gregory Chaitin, *Information, Randomness and Incompleteness—Papers on Algorithmic Information Theory*, second edition (Singapore: World Scientific Press, 1990).
8. The reader may be familiar with so-called chaotic systems. These are examples where the behavior of the system is essentially random, and no algorithmic compression is possible.
9. Some viruses use RNA rather than DNA (see chapter 5). This is one

example, taken from Bernd-Olaf Küppers, *Information and the Origin of Life* (Cambridge, Mass.: MIT Press, 1990), p. 101.

10. Most genomes will not be a completely random sequence, of course, if only because of the rules of punctuation in the genetic code. In addition, whole chunks of DNA may get duplicated or inverted, especially in eukaryotes. However, we can strip out these simple large-scale regularities and still ask whether what remains is random. Within individual protein-specifying sequences, no systematic patterning has been discerned.

11. Schrödinger was explicit about this when he conjectured that the genome must consist of "an aperiodic crystal." He drew the comparison between a normal crystal and a wallpaper pattern, remarking that the genome was more like a tapestry. See Erwin Schrödinger, *What Is Life?* (Cambridge: Cambridge University Press, 1944), p. 64. A very clear discussion of the distinction between order and organization, including a detailed account of why genomes are both random and specific, is given in Hubert Yockey, *Information Theory and Molecular Biology* (Cambridge: Cambridge University Press, 1992). I shall return to this topic in chapter 10.

12. Positive answers right down to the level of single-celled organisms have been given by Stuart Hameroff in "Quantum Coherence in Microtubules: A Neural Basis for Emergent Consciousness?," *Journal of Consciousness Studies* 1(1994):91.

CHAPTER 5: THE CHICKEN-AND-EGG PARADOX

1. See, for example, Michael Behe, *Darwin's Black Box* (New York: Free Press, 1996).

2. For a popular account, see Thomas Cech, "RNA as an Enzyme," *Scientific American* 255, no. 5 (1986):64.

3. Biologists use the word "genotype" to refer to the information in the genome that gets passed on between generations, and "phenotype" to refer to the actual expression of the genotype in the form of the living organism. In the RNA world, genotype and phenotype are one and the same.

4. For a review, see Sol Spiegelman, "An *In Vitro* Analysis of a Replicating Molecule," *American Scientist* 55(1967):221.

5. M. Eigen and P. Schuster, *The Hypercycle: The Principle of Natural Self-Organization* (Berlin: Springer-Verlag, 1979), pt. 2, chap. 14.

6. Ibid.
7. Of course, we have to explain why the world isn't full of these easy-to-make mini-replicators. One explanation might be that they do exist but inhabit a realm very different from that in which life now flourishes—e.g., the interior of a comet or the atmosphere of Titan, a moon of Saturn (see chapter 9). Another is that they are destroyed by organic life as soon as they form.
8. Julius Rebek, "Synthetic Self-Replicating Molecules," *Scientific American* 271, no. 1 (1994):34.
9. For a report, see Philip Cohen, "Can Protein Spring into Life?," *New Scientist*, April 26, 1997, p. 18. The work itself is published in *Nature* 382(1996):525.
10. Freeman Dyson, *Origins of Life* (Cambridge: Cambridge University Press, 1985).
11. E. G. Nisbet, *The Young Earth* (London: Allen & Unwin, 1987), chap. 8.
12. Michael Russell, Roy Daniel, Allan Hall, and John Sherringham, "A Hydrothermally Precipitated Catalytic Iron Sulphide Membrane as a First Step Toward Life," *Journal of Molecular Evolution* 39(1994):231. For a popular account, see Michael Russell, "Life from the Depths," *Science Spectra* 1(1996):26.
13. A popular account of his theory is given in A. G. Cairns-Smith, *Seven Clues to the Origin of Life* (Cambridge: Cambridge University Press, 1985).
14. Ilya Prigogine and Isabelle Stengers, *Order Out of Chaos* (London: Heinemann, 1984), chap. 5.
15. For a popular account, see Stuart Kauffman, *At Home in the Universe* (Oxford: Oxford University Press, 1995).
16. John Maynard Smith, "Life at the Edge of Chaos?," *New York Review of Books*, March 2, 1995, p. 28.
17. See, for example, Prigogine and Stengers, *Order Out of Chaos*, chap. 5.

CHAPTER 6: THE COSMIC CONNECTION

1. Titus Lucretius Carus, *De Rerum Natura*, trans. Alban Dewes Winspear (New York: Harbor Press, 1956), p. 89.
2. The reasoning behind this is simple. The total mass of the Earth's carbon divided by the mass of carbon in your body, although a big num-

ber, is nevertheless far smaller than the number of carbon atoms in your body.

3. A. D. Taylor, W. J. Baggaley, and D. I. Steel, "Discovery of Interstellar Dust Entering the Earth's Atmosphere," *Nature*, 380(1996):323.

4. Quoted in C. B. Cosmovici, S. Bowyer, and D. Werthimer, eds., *Astronomical and Biochemical Origins and the Search for Life in the Universe* (Bologna: Editrice Compositori, 1997), p. 106. For a comprehensive review of the role of comets, see Paul Thomas, Christopher Chyba, and Christopher McKay, eds., *Comets and the Origin and Evolution of Life* (New York: Springer-Verlag, 1997).

5. James Hutton, "Theory of the Earth," *Transactions of the Royal Society of Edinburgh* 1(1788):304.

6. Norman Sleep, Kevin Zahnle, James Kasting, and Harold Morowitz, "Annihilation of Ecosystems by Large Asteroid Impacts on the Early Earth," *Nature* 342(1989):139.

7. Kevin Maher and David Stephenson, "Impact Frustration of the Origin of Life," *Nature* 331(1988):612.

CHAPTER 7: SUPERBUGS

1. C. D. Parker, "The Corrosion of Concrete," *Australian Journal of Experimental Biology and Medical Science* 23(1945):81, 91. For a general review of superbugs, see Michael Madigan and Barry Marrs, "Extremophiles," *Scientific American* 276, no. 4 (1997):66; and John Postgate, *The Outer Reaches of Life* (Cambridge: Cambridge University Press, 1996).

2. Erasmus Darwin, *The Temple of Nature* (London: J. Johnson, 1793).

3. See Gregory Bock and Jamie Goode, eds., *Evolution of Hydrothermal Ecosystems on Earth (and Mars?)* (New York: Wiley, 1996), p. 37.

4. Actually, it may not be completely dark. There can be an eerie glow around the vents caused by some as yet ill-understood process. Some scientists have conjectured that photosynthesis might have started from this faint submarine light, rather than from sunlight. See Ruth Flanagan, "The Light at the Bottom of the Sea," *New Scientist*, December 13, 1997, p. 42.

5. Most of the organisms living near black smokers are indirectly dependent on sunlight, either by making use of dissolved oxygen (a byproduct of photosynthesis) or by eating organic scraps that descend

from the surface. Thirty years ago the biologist George Wald wrote: "It may form an interesting intellectual exercise to imagine ways in which life might arise, and having arisen might maintain itself, on a dark planet; but I doubt very much that this has ever happened, or that it can happen." See "Life and Light," *Scientific American* 201, no. 4(1959):92. However, Wald was wrong. Chemotrophs that are truly independent of surface life are known.

6. T. Gold, "The Deep, Hot Biosphere," *Proceedings of the National Academy of Science USA* 89(1992):6045.

7. In 1955, marine biologists discovered bacteria in sediments from the bottom of the Pacific. Based on their analysis, they proclaimed with pedantic confidence that the biosphere terminated precisely 7.47 meters down! See R. Y. Morita and C. E. Zobell, "Occurrence of Bacteria in Pelagic Sediments Collected During the Mid-Pacific Expedition," *Deep-Sea Research* 3(1955):66–73.

8. Lloyd Hamilton, "Aspects of Metallogenesis and Microorganisms in the Red Sea Region of Saudi Arabia," unpublished Ph.D. thesis, University of London, 1973.

9. Gold, "Deep, Hot Biosphere."

10. J. P. McKinley et al., "D.O.E. Seeks Origin of Deep Subsurface Bacteria," *EOS: Transactions of the American Geophysical Union* 75(1994): 385.

11. The idea of exploiting microbes to aid oil extraction was first mooted by a group of scientists in Australia in 1983, but their proposal fell on deaf ears. A published account can be found in B. Bubela, P. L. Stark, and M. Kords, "Microbiologically Enhanced Oil Recovery," in *Baas Becking Geobiological Laboratory Annual Report* (Canberra: Commonwealth of Australia, 1983), p. 53. More recently, several commercial organizations have begun taking a keen interest in the untapped bioresources that lie underground. In addition to improvements in the oil-and-gas sector, we could soon see the burgeoning of a major new industry called bioremediation—cleaning up polluted land and water using customized microbes. Trillions of dollars in clean-up costs could be saved by setting superbugs to work, eating toxic chemicals in deep or inaccessible sites. There is also a huge potential for identifying and exploiting unique enzymes and other molecular agents that confer such amazing capabilities on these organisms. The U.S. National Cancer Institute has already screened over five thousand cultures of

subsurface organisms in the search for anticancer drugs and AIDS vaccines.

12. Tim Appenzeller, "Deep-Living Microbes Mount a Relentless Attack on Rock," *Science* 258(1992):222.

13. R. J. Parkes et al., "Deep Bacterial Biosphere in Pacific Ocean Sediments," *Nature* 371(1994):410.

14. Everett Shock, "High Temperature Life Without Photosynthesis as a Model for Mars," *Journal of Geophysical Research—Planets* 102(1997): 23,687.

15. Ibid., p. 23,691.

16. More accurately, the evidence suggests that thermophiles evolve more slowly than low-temperature microbes. Since most archaea are thermophiles or hyperthermophiles, archaea as a class tend to have evolved less than bacteria. However, there are some hyperthermophilic bacteria, such as Aquifex, that have also evolved very slowly, whereas some mesophilic archaea have undergone substantial evolutionary changes. I am grateful to Susan Barns for drawing my attention to this.

17. See Bock and Goode, *Evolution of Hydrothermal Ecosystems*, chaps. 1, 2.

18. J. B. Corliss et al., "Submarine Thermal Springs on the Galápagos Rift," *Science* 203(1979):1073.

19. Gold, "Deep, Hot Biosphere." See also Karsten Pedersen, "The Deep Subterranean Biosphere," *Earth Science Reviews* 34(1993):243.

20. G. Wächterhäuser, "Evolution of the First Metabolic Cycles," *Proceedings of the National Academy of Science USA* 87(1990):200.

21. Todd Stevens and James McKinley, "Lithoautotrophic Microbial Ecosystems in Deep Basalt Aquifers," *Science* 270(1995):450. For popular accounts, see James Fredrickson and Tullis Onstott, "Microbes Deep Inside the Earth," *Scientific American* 275, no. 4 (1996):42; Larry O'Hanlon, "How Life Would Be at Home on Mars," *New Scientist*, October 28, 1995, p. 19; Stephanie Pain, "The Intraterrestrials," *New Scientist*, March 7, 1998, p. 28.

22. This theory might also explain the existence of extreme halophiles—archaea that live in very salty conditions. As the bombardment abated, Earth would nevertheless have continued to suffer impacts big enough to boil part of the oceans, producing layers of concentrated brine that would have been lethal to any organisms that were not both salt- and

heat-tolerant. However, the evidence is less compelling in this case, because most extant halophiles are not obviously of ancient lineage.

CHAPTER 8: MARS: RED AND DEAD?

1. Percival Lowell, *Mars and Its Canals* (New York: Macmillan, 1906), p.
   376.

2. There is a hint of effluent from still-active volcanic vents in the
   canyons of Valles Marineris. Also, the data from *Pathfinder* suggest
   very recent volcanic activity.

3. See, for example, Penelope Boston, Mikhail Ivanov, and Christopher
   McKay, "On the Possibility of Chemosynthetic Ecosystems in Subsurface Habitats on Mars," *Icarus* 95(1992):300. For a popular account by
   the same authors, see "Considering the Improbable: Life Underground
   on Mars," *Planetary Report* 14(1994):13.

4. Christopher McKay, "The Search for Life on Mars," *Origins of Life and
   Evolution of the Biosphere* 27(1997):263. For a popular account, see *Astronomy*, August 1997, p. 38.

5. D. D. Bogard, L. E. Nyquist, and P. Johnson, "Noble Gas Contents of
   Shergottites and Implications for the Martian Origin of SNC Meteorites," *Geochimica et Cosmochimica Acta* 48(1984):1723. For a popular
   account, see Andrew Chaikin, "A Stone's Throw from the Planets,"
   Sky and Telescope, February 1983, p. 122.

6. "Statement from Daniel S. Golding, NASA Administrator" (Johnson
   Space Center press release), *NASA News*, August 6, 1996, pp. 96–
   159.

7. David Mittlefehldt, "The Source of ALH84001," *Planetary Report*
   17(1997):5.

8. D. S. McKay et al., "Search for Past Life on Mars: Possible Relic Biogenic Activity in Martian Meteorite ALH84001," *Science* 273(1996):
   924. A recent review of the evidence for and against traces of life
   in ALH84001 is given in *Planetary Report* 18, no. 3, May/June 1998,
   p. 9.

9. Robert Folk and F. Leo Lynch, "The Possible Role of Nanobacteria
   (Dwarf Bacteria) in Clay Mineral Diagenesis, and the Importance of
   Sample Preparation in High Magnification SEM Study," *Journal of
   Sedimentary Research* 67(1997):583.

10. Olavi Kajander et al., "Nanobacteria from Blood, the Smallest Cultur-

able Autonomously Replicating Agent on Earth," *Society of Photo-optical Instrumentation (The International Society for Optical Engineering)* 3111(1997):420.

11. I. P. Wright, M. M. Grady, and C. T. Pillinger, "Organic Materials in a Martian Meteorite," *Nature* 340(1989):220.

12. Thomas Jukes, "Lessons from Evolution: Ruling Out Danger," *Planetary Report* 14(1994):14.

13. Cited in John Rummel and Michael Meyer, "Where No One Has Gone Before: What Is Planetary Protection Anyway?," *Planetary Report* 14(1994):5.

14. Kenneth Nealson et al., *Mars Sample Return: Issues and Recommendations* (Washington, D.C.: National Academy Press, 1997), p. 3.

15. Taped interview with Tim Radford of *The Guardian* (U.K.) in 1996.

16. Quoted in Ryder Miller, "The Natural Universe," *Mercury* 26(1997):28.

17. Quoted in Nealson et al., *Mars Sample Return,* p. 15.

18. *Guardian* interview.

19. Jack D. Farmer, "Exploring Mars for Evidence of Past or Present Life: Roles of Robotic and Human Missions," abstract of paper delivered at "Origins" conference, Estes Park, Colorado, May 1997.

20. This intuitive idea can be precisely quantified with something called Bayes' rule, often applied to evidence presented in a court of law. For example, suppose the accused is already judged very likely to be guilty, and additional fingerprint evidence is presented. The jury is told that the chances of an accidental fingerprint match are ten to one. This would be enough to convict the accused. On the other hand, if the accused looks very likely to be innocent, the fingerprint match is less significant. It is a fallacy to conclude in the case of a ten-to-one match that there is a 90 percent chance of guilt. The odds must be weighted by the prior probability of guilt before an overall likelihood can be computed. In the case of life on Mars, this prior probability varies dramatically—from very close indeed to zero, to something approaching one, depending on your assumption about panspermia (see chapter 9).

CHAPTER 9: PANSPERMIA

1. Svante Arrhenius, *Worlds in the Making* (London: Harper, 1908).

2. G. Horneck, H. Bucker, and G. Reitz, "Long-Term Survival of Bacterial Spores in Space," *Advanced Space Research* 14(1994):1041.

3. J. Koike et al., "Survival Rates of Some Terrestrial Microorganisms Under Simulated Space Conditions," *Advanced Space Research* 12(1992):4271.

4. Peter Weber and Mayo Greenberg, "Can Spores Survive in Interstellar Space?" *Nature* 316(1985):403. See also the follow-up work of Curt Mileikowsky, "Can Spores Survive a Million Years in the Radiation of Outer Space?," in C. B. Cosmovici, S. Bowyer, and D. Werthimer, eds., *Astronomical and Biochemical Origins and the Search for Life in the Universe* (Bologna: Editrice Compositori, 1997), p. 545.

5. Fred Hoyle, *The Intelligent Universe* (London: Michael Joseph, 1983).

6. Paul Wesson, Jeff Secker, and James Lepock, "Panspermia Revisited: Astrophysical and Biological Constraints," in Cosmovici et al., eds., *Astronomical and Biochemical Origins*, p. 539.

7. Only very recently have astronomers positively identified any planets outside our solar system. The problem about detecting extrasolar planets is that they are too faint to show up in even the most powerful telescope. Their existence can be inferred only indirectly. As a planet orbits its star, it exerts a gravitational tug, causing the star to execute a little wiggle. The effect is extremely small, but it can show up in the light spectrum of the star in a distinctive way. As a result of very careful observations, several large planets have been detected within a few light-years of Earth. Current techniques aren't good enough to spot a planet with a mass and orbit similar to the Earth's, but, given the existence of other planetary systems, it seems extremely likely that Earth-like planets are out there somewhere. There are probably many millions of them in our galaxy alone, each a potential abode for life. See Paul Halpern, *The Quest for Alien Planets* (London: Plenum Press, 1977).

8. George Claus and Bartholomew Nagy, "A Microbiological Examination of Some Carbonaceous Chondrites," *Nature* 192(1961):594. For an excellent review of the history of the search for life in meteorites, with special reference to the Murchison meteorite, see David Seargent, *Genesis Stone* (Sydney: Karagi Publications, 1991).

9. Evidence of life in the Murchison meteorite has been presented by the German scientist Hans Pflüg, who studied thin sections of the rock under an optical microscope. Pflüg found some very curious structures that look remarkably like filamentary bacteria. See "Ultrafine Structure of the Organic Matter in Meteorites," in N. C. Wickramasinghe, ed., *Fundamental Studies and the Future of Science* (Cardiff, Wales: Uni-

versity College Cardiff Press, 1984), p. 24. Other scientists have been quick to dismiss Pflüg's claim.

10. H. J. Melosh, "The Rocky Road to Panspermia," *Nature* 332(1988):687. For a popular account, see H. J. Melosh, "Swapping Rocks: Exchange of Surface Material Among the Planets," *Planetary Report* 14(1994):16.

11. Quoted in Arrhenius, *Worlds in the Making*, p. 219.

12. In 1964, the distinguished biologist George Gaylord Simpson wrote, "It is extremely improbable, almost to the point of impossibility, that any form of life has ever traveled by natural means from one planetary system to another," but he nevertheless concluded, "Such travel between earth and Mars, within the same planetary system, is still improbable, but the possibility is not absolutely ruled out." See George Gaylord Simpson, "On the Nonprevalence of Humanoids," *Science* 143(1964):772.

13. Brett Gladman et al., "The Exchange of Impact Ejecta Between Terrestrial Planets," *Science* 271(1996):1387.

14. California microbiologist Raul Clano has reportedly formed a company to screen these ancient microbes for possible use as agricultural control agents. This was reported by the New Zealand Press Agency in, for example, *The Weekend Australian*, April 12, 1997.

15. John Postgate, *The Outer Reaches of Life* (Cambridge: Cambridge University Press, 1994).

16. Kenneth Nealson et al., *Mars Sample Return: Issues and Recommendations* (Washington, D.C.: National Academy Press, 1997), p. 18.

17. See Christopher McKay, "The Search for Life on Mars," *Origins of Life and Evolution of the Biosphere* 27(1997):263.

18. Carl Sagan, "Is It Dangerous to Return Samples from Mars to Earth?," *Planetary Report* 14(1994):3.

19. T. Gold, "The Deep, Hot Biosphere," *Proceedings of the National Academy of Science USA* 89(1992):6048.

20. Christopher McKay, "Promethean Ice," *Mercury* 25(1996):15.

21. Fred Hoyle and Chandra Wickramasinghe, *Diseases from Space* (London: Dent, 1979).

CHAPTER 10: A BIO-FRIENDLY UNIVERSE?

1. Freeman Dyson, *Disturbing the Universe* (New York: Harper & Row, 1979), p. 250.

2. *Philadelphia Inquirer*, April 9, 1997.

3. Fred Hoyle and Chandra Wickramasinghe, *Lifecloud* (London: Dent, 1978).

4. Fred Hoyle, *The Intelligent Universe* (London: Michael Joseph, 1983).

5. Francis Crick, *Life Itself: Its Nature and Origin* (New York: Simon & Schuster, 1981).

6. Jacques Monod, *Chance and Necessity*, trans. A. Wainhouse (New York: Knopf, 1971), p. 145.

7. Christian de Duve, *Vital Dust: Life as a Cosmic Imperative* (New York: Basic Books, 1995), p. 300.

8. Kenneth Nealson et al., *Mars Sample Return: Issues and Recommendations* (Washington, D.C.: National Academy Press, 1997), p. 13.

9. For a sharply critical appraisal of this philosophy, see Robert Shapiro, *Origins: A Skeptic's Guide to the Creation of Life on Earth* (New York: Summit Books, 1986).

10. De Duve, *Vital Dust*, pp. xv, xviii.

11. As quoted in D. S. Bendall, *Evolution from Molecules to Men* (Cambridge: Cambridge University Press, 1983), p. 128.

12. Gary Steinman and Marian Cole, "Synthesis of Biologically Pertinent Peptides Under Possible Primordial Conditions," *Proceedings of the National Academy of Science* 58(1976):735.

13. Sidney Fox, "Prebiotic Roots of Informed Protein Synthesis," in Horst Kleinkauf, Hans von Dohren, and Lothar Jaenicke, eds., *The Roots of Modern Biochemistry* (Berlin: de Gruyter, 1988), p. 897.

14. Quoted in Shapiro, *Origins*, pp. 186–87.

15. Cyril Ponnamperuma, "The Origin, Evolution and Distribution of Life in the Universe," in Clifford Matthews and Roy Abraham Varghese, eds., *Cosmic Beginnings and Human Ends* (Chicago: Open Court, 1993), p. 91.

16. This result is related to Gödel's incompleteness theorem in mathematics.

17. Fractals are commercially exploited in data storage of complex images. It is cheaper in terms of information to fractalize rather than pixelate many natural images. See, for example, Barry Fox, "Fractals Set the Pattern for Online Video," *New Scientist*, September 7, 1996, p. 23.

18. See, for example, Paul Davies, *The Cosmic Blueprint* (London: Heinemann, 1987), chap. 5.

19. Christian de Duve, "The Chemical Origin of Life," in C. B. Cos-

movici, S. Bowyer, and D. Werthimer, eds., *Astronomical and Biochemical Origins and the Search for Life in the Universe* (Bologna: Editrice Compositori, 1997), p. 392.

20. Stuart Kauffman, *The Origins of Order* (Oxford: Oxford University Press, 1993), p. 285.
21. See, for example, Peter Coveney and Roger Highfield, *Frontiers of Complexity* (New York: Ballantine, 1995), chap. 4.
22. Christopher Langton, *Artificial Life* (Redwood City, Calif.: Addison-Wesley, 1988). For a popular account, see Coveney and Highfield, *Frontiers of Complexity*, chap. 8.
23. In chapter 1, I floated the idea that gravitation may have a part to play in this.
24. Manfred Eigen, *Steps Towards Life*, trans. P. Woolley (Oxford: Oxford University Press, 1992), p. 12.
25. For example, Küppers writes: "Therefore alongside the Darwinian principle there must be a further principle of self-organization of matter governing the transition from the non-living to the living." (Bernd-Olaf Küppers, *Molecular Theory of Evolution* [Berlin: Springer-Verlag, 1985], p. 279).
26. This scenario is similar to that described by Eigen, in which hypercycles amplify the selectivity of the system and can override molecular Darwinism (see chapter 5).
27. For a review of the quantum-measurement problem, see, for example, P. C. W. Davies and J. R. Brown, eds., *The Ghost in the Atom* (Cambridge: Cambridge University Press, 1986).
28. Roger Penrose, *The Emperor's New Mind* (Oxford: Oxford University Press, 1989), chap. 10.
29. Curiously, DNA has tenfold, hence fivefold, symmetry when viewed end on.
30. See Gerard Milburn, *The Feynman Processor* (Sydney: Allen & Unwin, 1988) and Julian Brown, *Minds, Machines and the Multiverse: The Quest for the Quantum Computer* (New York: Simon & Schuster, forthcoming).
31. De Duve, *Vital Dust*, p. xviii.
32. Stephen Jay Gould, *Life's Grandeur* (London: Jonathan Cape, 1996).
33. Ibid., pp. 202–12.
34. Richard Dawkins, *Climbing Mount Improbable* (London: Viking, 1996).

35. Gould, *Life's Grandeur*, p. 167.
36. Ibid., p. 216.
37. De Duve, *Vital Dust*, p. 299.
38. Allan Dressler, *Voyage to the Great Attractor* (New York: Knopf, 1994), p. 335.
39. George Gaylord Simpson, "On the Nonprevalence of Humanoids," *Science* 143(1964):772.
40. "The Search for Extraterrestrial Intelligence: Scientific Quest or Hopeful Folly?" (debate between Ernst Mayr and Carl Sagan), *Planetary Report* 16(1996):4.
41. Gould, *Life's Grandeur*, pp. 175, 214, 216.
42. Monod, *Chance and Necessity*, p. 180.

INDEX